Beyond Coal

The Rise of Renewables and the Nuclear Issue

Francisco José Hurtado Mayén

Content

Introduction 7
    Objective of the book 7
    Current State of Renewable Energy 11
        Brief History and Situation of Renewable Energies 11
        History of Renewable Energy 11
        Current Situation of the Renewable Energy Sector 13
    Challenges and opportunities 19
        Challenges of Renewable Energy Adoption 19
        Long-Term Opportunities and Benefits 22

Chapter 1: Advances in Solar Technologies 27
    Photovoltaic Technology 27
        Evolution and Improvements in Solar Panel Efficiency 27
        New Materials and Emerging Technologies 29
        Concentrated Solar Energy (CSP) 32
        Recent Innovations and Future CSP Developments 36
        Integration and Optimization 39

Chapter 2: Innovations in Wind Energy 45
    Onshore wind turbines 45
    Offshore Wind Energy 50
    Emerging Technologies 54

Chapter 3: Emerging Renewables 59
    Tidal Energy 59
    Wave Energy 62
    Other Energy Sources 66

Geothermal energy 67

Biomass 71

Other Innovative Sources 75

Potential and Future Prospects 78

Chapter 4: Innovations in Storage and Management 81

Advanced Batteries 81

Thermal and Mechanical Storage 85

Electricity Network Management 88

Chapter 5: Political and Economic Framework 93

Government Policies 93

Economic Incentives 97

Financing Strategies 102

Chapter 6: The Role of Artificial Intelligence and Big Data 109

Power System Optimization 109

Predictive Maintenance 113

Data Analysis and Decision Making 117

Chapter 7: The Nuclear Energy Debate 121

Introduction to Nuclear Energy 121

Nuclear Energy as Renewable Energy 124

Against Nuclear Energy as Renewable Energy 127

Global Perspectives on Nuclear Energy 131

The Future of Nuclear Energy 134

Conclusions on Nuclear Energy 138

Conclusion 143

In short 143

What about the future? 146

What can you do? 150
Appendices 153
    Frequently asked questions... 153
    Glossary of Technical Terms.. 163
    Recommended Reading List... 169
    Helpful Resources and Tools 175
        Online Resources.. 175
        Tools & Software 177
        Publications and Databases...................................... 179
        Professional Organizations and Networks 181

Introduction

Objective of the book

In recent decades, humanity has faced unprecedented challenges related to climate change, environmental degradation and energy sustainability. These challenges have driven a radical transformation in the energy sector, driven by the urgent need to reduce greenhouse gas emissions and become less dependent on fossil fuels. "Beyond Carbon" is conceived as a comprehensive and visionary guide that explores the frontiers of innovation in renewable energies and analyzes the controversial but hopeful role of nuclear energy in the energy mix of the future.

The main purpose of this book is to provide readers with an in-depth and up-to-date understanding of the most advanced technologies in the field of renewable energy, as well as the debate around nuclear energy. Through a combination of technical analysis, case studies, expert interviews, and public policy reflections, the book aims to shed light on how these innovations are transforming the global energy landscape and what the long-term prospects are.

This book is of utmost importance for several reasons:

- Education and Awareness: Provides detailed and accessible information on the most advanced and emerging technologies in the field of renewable energy, helping to educate and raise awareness among a wide audience. This book is aimed at professionals in the sector as well as students, researchers and the public interested in sustainability. By demystifying technical concepts and presenting information in a clear and understandable way, the book seeks to empower readers with the knowledge needed to understand and participate in the debate about the energy future.
- Fostering Informed Debate: By approaching nuclear energy from a balanced perspective, the book encourages an informed debate about its role in the energy future. Recognizing both the advances and benefits and the risks and controversies associated with nuclear energy, this book provides a platform for constructive dialogue. Diverse points of view are presented and the arguments of defenders and critics are analyzed, facilitating a comprehensive and nuanced understanding of the subject.

- Inspiration for Innovation: Presents innovations and success stories that can serve as inspiration for researchers, entrepreneurs, and policymakers. Through detailed case studies and interviews with pioneers in the field, the book highlights innovative projects and technological solutions that are changing the game in the energy sector. This section aims to inspire the next generation of innovators to continue exploring and developing new technologies that drive the transition to cleaner, more efficient energy.
- Decision-Making Guide: Provides analysis and reflections on the policies and economic frameworks that are facilitating the transition to a more sustainable energy future. It examines examples of successful policies in different parts of the world, as well as the challenges and opportunities faced by legislators and regulators. The book also discusses innovative business models and financing mechanisms that are helping to catalyze investments in renewable energy and energy storage technologies.
- Promoting Energy Sustainability and Resilience: In a global context of growing energy demand and concerns about energy security, the book addresses how renewables and nuclear energy can

contribute to a more resilient and sustainable energy matrix. The synergies between different energy sources are analyzed and strategies for effective integration into the electricity grid are proposed. This holistic approach seeks to ensure that the energy solutions of the future are not only efficient and clean, but also resilient in the face of change and crisis.

- Global Perspectives and Regional Adaptability: Through a series of international case studies, the book examines how different regions are adopting and innovating in the use of renewable and nuclear energy. The geographical, economic and political particularities that influence the adoption of these technologies in different contexts are addressed, providing a global vision that can be adapted to local realities.

In summary, "Beyond Carbon" not only aspires to be a reference resource in the field of renewable energy and nuclear energy, but also to inspire and equip its readers with the knowledge and vision necessary to actively contribute to a sustainable and resilient energy future. By exploring technological advances, the policy framework, and ethical and economic debates, this book seeks to provide a roadmap for a future in which clean and secure energy is accessible to all.

Current State of Renewable Energy

Brief History and Situation of Renewable Energies

The history of renewable energy dates back to ancient times, when civilizations began using natural resources such as wind, water, and the sun to meet their basic energy needs. However, it was in the modern era that the development and adoption of renewable energy saw significant growth, driven by the industrial revolution and, more recently, by growing concerns about climate change and environmental sustainability.

History of Renewable Energy

Since ancient times, humans have harnessed the energy of moving water for various applications. Watermills were used by the Greeks and Romans to grind grain and perform other mechanical tasks. These machines took advantage of the current of the rivers to convert the movement of water into mechanical energy. In China and the Middle East, similar techniques were also used for irrigation and milling, demonstrating the wide utilization and adaptability of hydropower in different cultures and geographies.

The earliest references to the use of wind date back more than 2,000 years, with Persian windmills being used to pump water and grind grain. These mills were rudimentary but effective structures that transformed the kinetic energy of the wind into usable mechanical energy. During the Middle Ages, windmills spread throughout Europe, especially in the Netherlands, where they became an icon of the landscape. Dutch windmills not only ground grain, but were also employed in land drainage and other agricultural uses.

During the first Industrial Revolution, the invention of the steam engine and the growing demand for coal temporarily eclipsed the use of renewable energy. However, hydropower remained an important source for powering factories and machinery, especially in areas where coal was scarce or expensive. In this century, the first ideas about the use of solar energy also began to develop. In 1839, Alexandre Edmond Becquerel discovered the photovoltaic effect, which laid the foundation for the development of photovoltaic solar energy. This crucial discovery proved that sunlight could be converted directly into electricity. In the late 19th century, scientists such as Charles Fritts and Wilhelm Hallwachs conducted key experiments that led to the first rudimentary solar devices, paving the way for future innovations in solar technology.

In the 20th century, the development of wind and solar energy accelerated significantly. In the 1970s, the oil crisis sparked renewed interest in alternative energy sources, due to the need to reduce dependence on fossil fuels and mitigate the impacts of energy crises. Technological advances in wind turbines and solar panels began to make their use viable on a larger scale. During this period, the first wind farms and commercial solar plants were established, marking a milestone in the adoption of renewable energy.

Starting in the 1980s, several governments began to implement policies to encourage the adoption of renewable energy, such as subsidies and feed-in tariffs. These policies helped reduce costs and increase investment in renewable technologies, making solar and wind more competitive compared to traditional energy sources. The construction of large hydroelectric dams in countries such as Brazil, with the Itaipu hydroelectric plant; China, with the monumental Three Gorges Dam; and the United States, with the iconic Hoover Dam, marked the rise of modern hydroelectric power. These projects not only provided large amounts of electricity, but also demonstrated the potential of hydropower to contribute significantly to the global energy matrix, cementing it as one of the leading sources of renewable energy globally.

Current Situation of the Renewable Energy Sector

Today, the renewable energy sector is at the heart of the global energy transition. Technological advancements, government support, and growing demand for clean energy have led to rapid growth in this sector. The installed capacity of solar PV has grown exponentially over the past two decades, with countries such as China, the United States and Germany leading the adoption of this technology. Global solar power capacity has surpassed 700 gigawatts (GW) in 2023 and is expected to continue to grow as costs continue to decline. The costs of solar panels have decreased significantly, making them more accessible and competitive compared to traditional energy sources. In the last decade, solar energy costs have been reduced by more than 80%, thanks to improvements in production technology and economies of scale. Solar panel technologies have advanced, including perovskite solar cells, concentrating solar systems (CSPs), and bifacial panels that capture sunlight from both sides. These innovations are improving the efficiency and cost-effectiveness of solar energy.

Wind energy has seen a global expansion, with large wind farms both onshore and offshore. Denmark, the United States and China are leaders in this technology. Global installed wind power capacity has exceeded 750 GW, with increasing investment in offshore projects. Wind turbines have evolved to be more efficient and capable of generating more electricity from wind, even in moderate wind conditions. Modern wind turbines can have blades that are more than 100 meters long and towers that exceed 150 meters in height, significantly increasing the generation capacity. Offshore wind is emerging as a crucial component of renewable energy. Countries such as the United Kingdom and Germany have developed extensive offshore facilities, taking advantage of strong and constant offshore winds to generate electricity efficiently.

Hydropower remains one of the largest sources of renewable energy, especially in water-rich countries such as Brazil, Canada, and Norway. It accounts for approximately 16% of global electricity production. Although it is a source of clean energy, the construction of large hydroelectric dams can have significant environmental impacts, such as the alteration of aquatic ecosystems and the displacement of communities. This has led to increased scrutiny and the search for more sustainable solutions, such as run-of-the-river hydropower and the modernization of existing infrastructure. Small-scale hydropower is gaining popularity as a less disruptive alternative. These projects have a lower environmental impact and can be implemented in rural and remote communities.

Emerging renewables, such as biomass and biogas, are being used to generate electricity and heat, especially in rural and agricultural areas. Biomass can come from agricultural, forestry, and urban waste, while biogas is produced from the anaerobic decomposition of organic matter. Geothermal energy, used primarily in regions with geothermal activity, such as Iceland and some U.S. states, provides a consistent and reliable source of energy. Global installed geothermal power capacity is growing, with new projects in countries such as Kenya, the Philippines and Indonesia. Technologies such as wave energy and tidal energy are in the development and demonstration phases. Pilot projects in places such as the UK, Canada and Australia are exploring the potential of these sources to contribute significantly to the energy mix.

The innovations and future of the renewable energy sector are promising. The integration of energy storage systems, such as lithium-ion batteries, is enabling greater stability and reliability in the supply of renewable energy. Large-scale storage is making it easier to manage the intermittency of sources such as solar and wind. The implementation of smart grids is improving energy management and distribution, facilitating greater penetration of renewable energies in the energy mix. These grids use advanced technologies to monitor and manage the flow of electricity efficiently, adapting to fluctuations in supply and demand. Hydrogen production from renewables is gaining attention as a potential solution for long-term energy storage and decarbonization of hard-to-electrify sectors such as heavy industry and long-distance transportation.

Historically, renewable energy has gone from being a curiosity to an urgent need in the current context of climate change and the search for sustainability. From ancient water and windmills to modern solar and wind technologies, progress has been remarkable. For centuries, humanity has been looking for ways to harness the forces of nature to perform everyday tasks. The industrial revolution, although dominated by coal and steam, could not completely eclipse the potential of renewable energies. In the 19th century, the invention of the steam engine and the growing demand for coal temporarily eclipsed the use of renewable energy. However, hydropower remained an important source for powering factories and machinery. In this century, the first ideas about the use of solar energy also began to develop. In 1839, Alexandre Edmond Becquerel discovered the photovoltaic effect, which laid the foundation for the development of photovoltaic solar energy. In the late 19th century, scientists such as Charles Fritts and Wilhelm Hallwachs conducted key experiments that led to the first rudimentary solar devices.

In the 20th century, the development of wind and solar energy accelerated significantly. In the 1970s, the oil crisis sparked renewed interest in alternative energy sources, due to the need to reduce dependence on fossil fuels and mitigate the impacts of energy crises. Technological advances in wind turbines and solar panels began to make their use viable on a larger scale. During this period, the first wind farms and commercial solar plants were established, marking a milestone in the adoption of renewable energy.

Starting in the 1980s, several governments began to implement policies to encourage the adoption of renewable energy, such as subsidies and feed-in tariffs. These policies helped reduce costs and increase investment in renewable technologies, making solar and wind more competitive compared to traditional energy sources. The construction of large hydroelectric dams in countries such as Brazil, with the Itaipu hydroelectric plant; China, with the monumental Three Gorges Dam; and the United States, with the iconic Hoover Dam, marked the rise of modern hydroelectric power. These projects not only provided large amounts of electricity, but also demonstrated the potential of hydropower to contribute significantly to the global energy matrix, cementing it as one of the leading sources of renewable energy globally.

Progress in the renewable energy sector has not only been technical, but also social and political. Growing awareness of climate change and its impacts has led to an increased demand for sustainable energy policies. Innovations in renewable technologies have been driven by the need to find solutions to global problems, such as reducing greenhouse gas emissions and finding safe and sustainable energy sources. Investment in research and development, coupled with favorable policies and growing public awareness, has led to mass adoption of these technologies. As we continue to innovate and refine these energy sources, the prospect of a sustainable energy future becomes increasingly achievable. With a focus on sustainability and efficiency, renewables are not only transforming the energy sector, but also providing key solutions for a cleaner and healthier world.

Challenges and opportunities

Challenges of Renewable Energy Adoption

Despite significant advances in the development and implementation of renewable energy technologies, there are still several challenges that need to be overcome to facilitate a global energy transition. The intermittency and reliability of renewable energy sources, such as solar and wind, represent a considerable obstacle. These energy sources are inherently intermittent because they depend on weather conditions that are not always predictable or constant. This poses a challenge to ensure a continuous and stable supply of electricity. To mitigate this problem, the implementation of large-scale energy storage systems, such as lithium-ion batteries and emerging technologies such as compressed air energy storage, is crucial. In addition, the diversification of the energy mix and the development of smart grids help to balance supply and demand, providing greater stability and reliability to the energy system.

Infrastructure and grid capacity are also significant challenges to renewable energy adoption. Today's transmission and distribution infrastructure is not always prepared to handle the massive integration of renewables. Traditional power grids may need significant retrofits to accommodate the variability and decentralized distribution of renewable energy sources. Investments in infrastructure modernization, network expansion, and smart grid development are essential. These improvements allow for more efficient energy management and a greater capacity to integrate renewable sources into the grid, thus facilitating a smoother transition to a sustainable energy system.

Another major hurdle is start-up costs and financing. Although the costs of renewable energy technologies have fallen considerably, initial projects may require significant capital investments. This can be a hurdle, especially in regions with financial constraints. However, there are solutions to facilitate the mobilization of capital for renewable energy projects. Innovative financing mechanisms, such as power purchase agreements (PPAs), public-private partnerships, and green energy investment funds, can play a crucial role. In addition, government subsidies and tax incentives can reduce the initial financial burden, making renewable energy projects more attractive and viable.

Social acceptance and government policies also play a crucial role in the adoption of renewable energy. Sometimes, this adoption faces resistance from local communities due to concerns about environmental impact, land use, and other social factors. In addition, inconsistency in government policies can create uncertainty and discourage investment. To improve social acceptance, community engagement and education are essential. Consistent and supportive government policies, along with the creation of clear and stable regulatory frameworks, can provide the necessary certainty for investors and businesses, thereby fostering an enabling environment for renewable energy growth.

Finally, the technical and logistical challenges cannot be underestimated. Deploying and maintaining renewable energy technologies in remote or inhospitable areas can be technically challenging and logistically complicated. However, technological innovations, such as modular systems and off-grid solutions, can facilitate implementation in these areas. In addition, training and local capacity building can help overcome logistical challenges, ensuring that communities can maintain and operate their renewable energy systems effectively.

In summary, although the transition to a renewable energy-based energy system presents several challenges, there are viable solutions and strategies that can facilitate this process. The implementation of advanced energy storage technologies, the modernization of infrastructure, the development of innovative financing mechanisms, the promotion of consistent and favorable government policies, and the adoption of technological and logistical innovations are crucial steps. With a coordinated and collaborative approach, it is possible to overcome these obstacles and move towards a more sustainable and resilient energy future, benefiting both the environment and global communities.

Long-Term Opportunities and Benefits

Despite the challenges, the adoption of renewable energy presents numerous opportunities and long-term benefits that can transform the global economy and improve the quality of life. The transition to renewable energy is crucial to reducing greenhouse gas emissions and combating climate change. Generating electricity from renewable sources emits significantly less CO_2 compared to fossil fuels. This change not only contributes to climate change mitigation, but also protects ecosystems, reduces the frequency and severity of extreme weather events, and improves global public health by reducing air pollution.

The adoption of renewable energies can also increase energy independence and the security of countries. By reducing dependence on fossil fuel imports, nations can better protect themselves against volatile oil and gas prices and minimize the risk of geopolitical conflicts related to energy resources. This energy independence provides greater economic and political stability, allowing countries to better manage their resources and focus on sustainable development.

The renewable energy sector is a significant source of job creation, generating opportunities from research and development to the manufacture, installation and maintenance of clean technologies. The expansion of renewables can boost economic development, especially in rural and remote areas, providing local jobs and long-term economic opportunities. This not only improves the local economy, but also promotes social inclusion and reduces migration to cities in search of employment.

Investment in renewable energy encourages innovation and the development of new technologies, which may have applications beyond the energy sector. Technological innovation can improve energy efficiency, reduce production costs and open up new business opportunities in various sectors. In addition, advancement in renewable technologies can spur industrial competitiveness and position nations at the forefront of the global economy.

Power generation from renewable sources significantly reduces air pollution compared to burning fossil fuels. This improvement in air quality has a direct impact on public health, reducing the incidence of respiratory and cardiovascular diseases and improving people's quality of life. Reducing pollution contributes to healthier environments, which translates into lower health care costs and higher labor productivity.

Renewable energies are fundamental for sustainable development, providing a clean and continuous source of energy that does not deplete natural resources. This approach ensures that the energy needs of present and future generations can be met without compromising the environment and available resources. Sustainable development promotes a balance between economic growth, environmental protection and social equity, ensuring a viable future for all.

Diversifying the energy matrix with renewable sources increases resilience to the impacts of climate change and other extreme events. A more resilient and adaptable energy infrastructure is crucial for maintaining energy supply in crisis situations and for post-disaster recovery. Renewable energies, by not depending on a constant supply of fuels, offer greater security in energy supply during emergency situations, contributing to stability and economic recovery.

In conclusion, although the adoption of renewable energy faces several challenges, the long-term opportunities and benefits are immense. The transition to an energy system based on clean and sustainable sources is not only an environmental necessity, but also an economic and social opportunity that can transform the global future of energy and the quality of life on our planet. The adoption of renewable energy is not only a response to today's environmental challenges, but also a comprehensive strategy to foster economic development, technological innovation, and community resilience. With a purposeful and collaborative approach, the energy transition can usher in a new era of global prosperity and sustainability.

Chapter 1: Advances in Solar Technologies

Photovoltaic Technology

Evolution and Improvements in Solar Panel Efficiency

Photovoltaic technology has come a long way since its discovery in the 19th century to becoming one of the leading sources of renewable energy today. The evolution of solar panels has been marked by constant improvements in efficiency, cost reductions and technological advances that have increased their commercial viability. The journey began with the discovery of the photovoltaic effect by Alexandre Edmond Becquerel in 1839, which laid the foundation for the development of solar technology. Despite this early discovery, the first practical solar cells were not developed until the 20th century. In 1954, Bell Labs developed the first silicon solar cell capable of converting sunlight into electricity with an efficiency of 6%. These early solar cells were expensive and were mainly used in space applications due to their high cost and low efficiency.

The oil crisis of the 1970s was a significant catalyst for research and development in alternative energies, including photovoltaics. This global energy crisis spurred more investment in the search for sustainable and safe energy sources, and solar technology began to gain more attention. During this period, the first commercial applications of solar panels began to emerge, mainly in remote and off-grid locations, where conventional electricity was inaccessible or too expensive. The research focused on improving the efficiency of solar cells and reducing production costs to make them more viable for general use.

In the 1980s, government support and subsidies played a crucial role in the expansion of research in photovoltaic technology. Governments around the world recognized the potential of solar energy and began offering incentives to stimulate its development. This support enabled significant advances, including the development of polycrystalline silicon panels, which while offering slightly lower efficiency than monocrystalline ones, were much cheaper to manufacture. These panels began to become more common in commercial and residential applications, expanding the adoption of solar energy.

The 2000s marked a period of cost reduction and increased efficiency in solar technology. Advances in manufacturing technology, coupled with economies of scale, resulted in significant cost reductions for solar panels. Between 2000 and 2020, solar panel costs decreased by more than 80%, making solar energy more accessible to a greater number of people and businesses. Continuous improvements in solar cell technology, including the use of techniques such as anti-reflective coating and post-contact structures, increased the efficiency of commercial solar panels to more than 20%. These advances allowed solar systems to generate more electricity per square meter, improving their economic and environmental viability.

In the 2020s, PV technology has reached new levels of sophistication with the introduction of advanced technologies such as heterojunction (HJT) and passivated back cell (PERC). These innovations have enabled even greater efficiencies, bringing solar cell efficiencies closer to their theoretical limit. Bifacial solar panels, which can capture sunlight from both sides, are gaining popularity, increasing energy production per square meter and improving the cost-effectiveness of solar installations. In addition, the integration of energy storage technologies, such as lithium-ion batteries, has improved the ability of solar systems to provide constant and reliable electricity, even when the sun is not shining.

New Materials and Emerging Technologies

In addition to improvements in traditional silicon technologies, research is exploring new materials and emerging technologies that promise to revolutionize the field of solar PV. Perovskite solar cells have emerged as a promising technology due to their high efficiency potential and low cost of production. In just over a decade of research, perovskite solar cells have achieved efficiencies more than 25% in the laboratory. This rapid advancement has captured the attention of scientists and industry, as perovskites can be manufactured at lower temperatures and with solution deposition techniques, which significantly reduces production costs. In addition, they could be applied on flexible substrates, opening possibilities for novel applications such as portable devices and curved surfaces.

However, despite their great potential, perovskite solar cells face significant challenges in terms of stability and durability. Exposure to moisture, oxygen, and UV light can quickly degrade these materials, limiting their lifespan. Current research focuses on improving degradation resistance and encapsulation to increase its durability. Scientists are developing new encapsulation techniques and materials that can protect perovskites from the elements, thereby extending their shelf life and making them a viable option for long-term commercial applications.

Third-generation technologies are also at the heart of solar research. Organic solar cells (OPVs) use conductive polymers and organic materials to absorb sunlight. They are flexible, lightweight, and can be produced at low cost, making them attractive for a variety of applications. However, their efficiency and durability are still inferior to those of silicon-based technologies, limiting their large-scale adoption. Researchers are working to improve the efficiency of OPVs and increase their lifespan by developing new materials and manufacturing techniques.

Thin-film solar cells, such as cadmium telluride (CdTe) and copper-indium gallium selenide (CIGS), offer the advantage of being lightweight and flexible, making them suitable for large, architectural surface applications. These cells have achieved competitive efficiencies and are particularly useful in situations where weight and flexibility are crucial. Nanoparticle and quantum dot solar cells are in the early stages of development, but they promise to improve efficiency and reduce costs by manipulating electronic properties at the nanometer scale. These emerging technologies have the potential to revolutionize the field of solar energy, but they still require a lot of research and development before they can be commercially viable.

Another promising area of photovoltaic technology is tandem solar cells. These cells combine two or more materials with different absorption bands to capture a larger portion of the solar spectrum. Combinations can include perovskites with silicon or with other perovskites, achieving efficiencies of more than 30% in the laboratory. However, integrating different materials and managing the interconnection between them are key challenges. Advances in materials engineering and manufacturing techniques are paving the way for the commercialization of these high-efficiency technologies. Tandem solar cells could provide a significant leap in solar energy conversion efficiency, making them a very active and promising area of research.

Advances in structures and design are also improving the overall efficiency of photovoltaic systems. The integration of microinverters and power optimizers allows for optimal performance in partial shading conditions and variability, improving the efficiency and energy production of solar systems. Solar tracking systems, which adjust the position of panels to follow the sun's path, are increasing energy production by up to 30%. These systems are especially effective in large installations and solar farms, where the increase in energy production may justify the additional cost of installing solar trackers.

Concentrated Solar Energy (CSP)

Concentrating Solar Energy (CSP) is an innovative technology that uses mirrors or lenses to concentrate a large area of sunlight onto a small receiver. This concentrated solar energy is converted into heat, which is then used to generate electricity through a thermal machine, usually a steam turbine, or for industrial processes that require heat. CSP systems harness the sun's abundant energy and concentrate it to produce electricity efficiently and sustainably.

The working principle of CSP systems is based on the concentration of sunlight. Using parabolic mirrors, solar towers, or linear reflectors, these systems focus sunlight on a specific receiver. This receiver absorbs solar energy and converts it into heat. Once sunlight has been concentrated and converted into heat, this heat is used to heat a working fluid, which can be water, molten salts, or synthetic oil. The hot fluid produces steam, which drives a turbine connected to an electric generator. One of the most prominent advantages of CSP systems is their ability to incorporate thermal storage. This storage allows electricity to be generated even when the sun is not shining, using the heat stored in materials such as molten salts. This ensures a constant supply of electricity at night or on cloudy days, increasing the reliability and efficiency of the system.

The current applications of CSP technology are varied and are implemented in different types of plants and systems. Solar tower plants, for example, use heliostats, which are flat or slightly curved mirrors, to follow the sun and concentrate its light on top of a central tower. A receiver in the tower collects the heat, which is then used to generate electricity. A prominent example of this technology is the Ivanpah Solar Plant in California, one of the largest CSP facilities in the world. This plant uses thousands of heliostats to concentrate sunlight onto three towers, producing enough electricity to power thousands of homes.

Another common application is the parabolic reflector plant. These systems use parabolic mirrors arranged in rows that concentrate sunlight into receiver tubes located at the focus of the mirrors. The heat collected in the tubes is used to generate steam and produce electricity. The Andasol Solar Plant in Spain is a notable example of this technology. With a capacity of 150 MW, this plant uses molten salt thermal storage to provide electricity even when there is no sun, thus improving the stability of the power grid.

Parabolic dish systems are another interesting CSP technology. They use dish-shaped parabolic mirrors to concentrate sunlight onto a receiver mounted on the focal point of the dish. The heat generated is used to drive a Stirling engine or microturbine, which produces electricity. These systems are known for their high efficiency and ability to operate autonomously, making them ideal for decentralized applications or in remote areas.

Parabolic trough plants are similar to parabolic reflectors but use curved channel-shaped mirrors to concentrate sunlight into receiver tubes that contain a heat transfer fluid. These plants are widely used in industrial applications to generate process steam or electricity. Parabolic trough technology is particularly efficient and can be deployed in a variety of industrial settings, providing a clean and sustainable source of energy.

CSP technology has advanced significantly over the past few decades, with continuous improvements in efficiency and reductions in costs. Research and development in this field have led to the implementation of more efficient and reliable systems, capable of providing a considerable amount of clean energy and reducing dependence on fossil fuels. Current CSP applications not only demonstrate its technological viability, but also its potential to contribute significantly to the global energy transition. With continued investment and support in research and development, CSP technology is well positioned to play a crucial role in the future of the world's energy supply, providing a sustainable and efficient energy source that can meet the growing energy needs of the world's population.

Recent Innovations and Future CSP Developments

As Concentrated Solar Power (CSP) technology matures, innovations continue to emerge that improve its efficiency, reduce costs, and expand its applications. These innovations are driven by the need to make solar energy more accessible and efficient, adapting to the growing demands for clean energy around the world. Advanced materials for high-temperature receivers are improving the thermal conversion efficiency and durability of CSP systems. Research into new materials, such as ceramics and selective coatings, is designed to withstand extreme temperatures and reduce heat losses, resulting in greater efficiency and receiver life.

The design of mirrors has also seen significant advances. Next-generation mirrors, made of lighter materials and advanced production techniques, are improving the accuracy of solar concentration and reducing installation and maintenance costs. These mirrors, coated with highly reflective and durable materials, are increasing the optical efficiency of CSP systems, allowing greater sunlight capture and, therefore, greater heat and electricity generation.

One of the most promising developments in the field of CSP is the advancement in thermal storage. The use of molten salts as a thermal storage medium is becoming a popular choice due to its high storage capacity and ability to operate at elevated temperatures. This allows for greater efficiency in electricity generation and longer storage capacity, which is crucial for providing constant power even when the sun is not shining. In addition, research in advanced thermal storage includes the development of phase-change materials (PCMs) and thermochemical storage technologies, which have the potential to increase storage density and improve thermal efficiency.

The integration of CSP with other technologies is also opening new opportunities. The combination of CSP systems with solar photovoltaic (PV) plants makes it possible to take advantage of both technologies. While PV systems generate electricity directly from sunlight, CSP systems with thermal storage can provide power during periods when PV generation declines, creating a more balanced and reliable energy supply. In addition, some developments are exploring the integration of CSP with fossil power plants to improve overall efficiency and reduce emissions. In these hybrid systems, the heat generated by CSP can be used to supplement the heat from fossil fuels, reducing dependence on conventional fuels and decreasing the carbon footprint.

Developments in small-scale CSP systems are making it possible to deploy them in decentralized and off-grid applications. These systems can provide electricity and heat to rural communities, industrial facilities, and other users in remote locations, where access to conventional energy is limited or expensive. In addition, the utilization of CSP for industrial processes that require high-temperature heat, such as desalination, hydrogen production, and the chemical industry, is expanding the applications of this technology beyond electricity generation. These industrial applications are proving that CSP is not only viable for energy production, but also as a sustainable heat source for various industries.

Artificial intelligence (AI) and advanced control systems are playing a crucial role in optimizing the performance of CSP plants. AI integration allows you to improve the accuracy of solar tracking, forecast system performance, and manage thermal storage more efficiently. AI algorithms can analyze large amounts of data in real-time to adjust mirrors and optimize sunlight harvesting, maximizing power generation. In addition, the use of data-driven predictive maintenance techniques is improving the reliability and reducing operating costs of CSP plants. Sensors and real-time data analysis enable early detection of failures and optimization of maintenance, ensuring that CSP plants operate efficiently and with minimal downtime.

In the future, these innovations are expected to continue to advance, making CSP technology even more efficient and accessible. As costs decrease and efficiency improves, CSP has the potential to play a crucial role in the transition to a clean and resilient energy economy. The combination of advanced materials, efficient thermal storage, and artificial intelligence technologies is positioning CSP as a key technology to meet the energy challenges of the future and contribute significantly to emissions reduction and global sustainability.

Integration and Optimization

The integration of solar energy into the electricity grid is essential to maximize its potential and ensure a stable and reliable electricity supply. As the share of solar energy in the global energy mix increases, various methods and strategies have been developed to facilitate its incorporation into existing electricity grids. Smart investors play a crucial role in this process. These devices convert the direct current (DC) generated by the solar panels into alternating current (AC) compatible with the power grid. In addition to performing this conversion, advanced smart inverters regulate frequency and voltage and provide ancillary services to the grid. They can respond quickly to fluctuations in solar power generation, helping to maintain grid stability. They also allow real-time monitoring and remote control, improving system management and performance.

Energy storage systems are another key piece in the integration of solar energy. Lithium-ion batteries, for example, make it possible to store excess energy generated during peak sunshine hours and release it when demand is high or solar generation is low. This not only improves grid reliability and stability, but also allows for greater solar energy penetration. Energy storage can provide ancillary services such as frequency control and backup support, contributing to a more robust and adaptable electrical system.

Smart grids represent a significant evolution in energy management. These networks use advanced communication and automation technologies to improve the efficiency, reliability and sustainability of the electricity supply. They can actively manage energy generation and consumption, as well as integrate distributed energy sources such as solar. Smart grids allow for dynamic, real-time management of solar energy, optimizing its use and minimizing the impact of intermittency. They also facilitate rapid detection and response to faults, improving grid resilience and ensuring a constant supply of electricity.

Microgrids are standalone energy systems that can operate autonomously or connect to the main grid. They include local power generation, such as solar, storage, and smart control. Microgrids improve the reliability of electricity supply in remote or critical areas and allow for greater integration of solar energy at the local level. They can also function as a backup in the event of mains failures, providing a flexible and resilient solution for energy management.

Demand prediction and management models use advanced weather data and algorithms to forecast solar generation and manage electricity demand efficiently. Accurate prediction of solar generation and active demand management help balance supply and demand, reducing the need for backup power and improving grid stability. These models make it possible to anticipate variations in power generation and adjust consumption, accordingly, optimizing the use of available resources.

Optimizing the integration of solar energy into the power grid is supported by advanced software tools and management techniques that improve the efficiency and reliability of the energy system. Energy management systems (EMS) monitor, control, and optimize the performance of solar energy systems and their interaction with the grid. EMS enables real-time management of power generation and consumption, improving operational efficiency and reducing costs. It also facilitates the integration of energy storage and demand response, ensuring optimal use of the energy generated.

Distributed grid management platforms (DERMS) coordinate and optimize the use of multiple distributed energy sources, including solar installations, energy storage, and other resources. DERMS improves visibility and control of distributed resources, optimizing their use and ensuring harmonious integration into the network. This facilitates rapid response to variations in supply and demand, while maintaining the stability of the energy system.

Optimization and simulation models use advanced algorithms to plan and manage the operation of solar energy systems and their integration into the grid. These models can identify the optimal strategies to maximize the efficiency and profitability of solar systems, as well as minimize integration and operating costs. Simulations make it possible to predict and mitigate potential impacts on the grid, ensuring that solar systems operate efficiently and without interruptions.

Data monitoring and analytics technologies collect and analyze real-time information about solar system performance and grid health. Detailed data collection and analysis allow for better decision-making and proactive management of solar energy systems. The data can be used to identify trends, optimize maintenance, and improve system reliability, ensuring optimal and long-lasting performance.

Artificial intelligence (AI) and machine learning (ML) are applied to optimize solar energy management, improve generation and consumption predictions, and develop advanced control strategies. AI and ML enable continuous and dynamic adaptation to changing conditions, improving the efficiency and stability of the energy system. These technologies can also identify patterns and anomalies, improving resilience and response to failures, contributing to a more robust and efficient energy system.

Advanced control algorithms manage solar energy generation, storage, and consumption to maximize system efficiency and reliability. Control algorithms can automatically balance supply and demand, optimize the use of energy storage, and coordinate demand response, improving system stability and cost-effectiveness. These algorithms ensure that solar energy is used effectively, minimizing waste and maximizing benefits for the power grid and consumers.

The integration and optimization of solar energy into the electricity grid are complex but essential processes for the development of a sustainable and efficient energy system. Through advanced integration methods, software tools, and optimization techniques, it is possible to effectively manage the intermittency of solar energy and make the most of its benefits. With the continued evolution of these technologies, solar energy will continue to play a key role in the transition to a sustainable energy future. The combination of technological innovation, strategic planning and efficient management is transforming the way solar energy is integrated into the grid, offering advanced and sustainable solutions to the energy challenges of the future.

Chapter 2: Innovations in Wind Energy

Onshore wind turbines

Onshore wind turbines have evolved significantly since their first designs, thanks to technological advances that have improved their efficiency, reliability and generation capacity. These advances have allowed wind energy to become one of the leading sources of renewable energy worldwide. One of the most important advances has been the design of blades. Advances in composite materials, such as carbon fiber and fiberglass, have allowed for the creation of longer and lighter blades. Longer blades can capture more energy from the wind, increasing the generating capacity of the turbines. In addition, improvements in the aerodynamics of the blades have reduced drag and increased efficiency. The blade shapes have been optimized using advanced modeling and simulation techniques to maximize energy capture and minimize wear.

The height of the masts has also undergone significant improvements. Wind turbine towers have been made taller to access stronger, more constant wind currents found at higher altitudes. Higher towers also allow for the use of longer blades, further improving efficiency. The use of high-strength materials and advanced construction techniques has allowed the construction of taller and more robust towers, capable of withstanding the additional loads. These advances have allowed onshore wind turbines to be more productive and efficient in generating electricity.

Generator technology has also advanced considerably. Permanent magnet generators have improved the efficiency and reliability of turbines by eliminating the need for an external excitation system and reducing mechanical and electrical losses. In addition, direct drive systems, which eliminate the gearbox, have improved reliability and reduced maintenance costs. These systems are also quieter and have a longer lifespan, making them more attractive to communities near wind farms.

The control and monitoring of wind turbines has improved with the implementation of advanced systems. Advanced control systems use sophisticated algorithms to optimize blade angle and turbine orientation in real-time, maximizing energy capture and protecting the turbine from harsh conditions. In addition, real-time monitoring systems collect data on turbine performance and condition, enabling proactive management and predictive maintenance. This reduces downtime and improves operational efficiency, ensuring that turbines are operating optimally.

Reducing environmental impact is another crucial aspect in the design of onshore wind turbines. Innovations in blade design have reduced the noise generated by the turbines, making them more acceptable to nearby communities. In addition, technologies and strategies have been developed to minimize the impact of turbines on wildlife, such as bird and bat detection and deterrence systems. These efforts have helped balance the need for renewable energy with environmental protection.

The successful implementation of onshore wind turbines in various parts of the world has demonstrated their viability and effectiveness as a renewable energy source. The Gansu Wind Farm in China, for example, is one of the largest wind projects in the world, with an installed capacity of more than 10 GW. This project has contributed significantly to the reduction of carbon emissions in the region and has helped diversify China's energy matrix. In addition, it has boosted local economic development and job creation.

The Alta Wind Energy Center (AWEC) in California, United States, is another notable example. Located in Tehachapi, this wind farm is one of the largest onshore in the world, with an installed capacity of approximately 1.5 GW. AWEC has been instrumental in providing clean energy to California's electric grid, helping the state achieve its renewable energy goals. The project has also generated employment and provided economic benefits to the local community.

In Australia, the Hornsdale Wind Farm is known for its combination with the Hornsdale Power Reserve, the world's largest lithium-ion battery, built by Tesla. With an installed capacity of 315 MW, this project has improved the reliability and stability of the electricity grid in South Australia. The combination of wind generation and battery storage has reduced carbon emissions and helped mitigate power outages in the region.

Another prominent project in the United States is Alta Farms, located in Illinois, with an installed capacity of 200 MW. This project uses advanced wind turbines that maximize energy capture in a region with moderate winds. Alta Farms has provided clean, renewable energy to thousands of homes in Illinois, contributing to the reduction of carbon emissions and meeting the state's renewable energy goals.

The Roscoe Wind Farm, located in West Texas, is one of the largest wind farms in the world, with an installed capacity of 781.5 MW. This project consists of more than 600 wind turbines distributed over a wide area. Roscoe has been a pioneer in integrating large-scale wind energy into the Texas power grid. The project has helped stabilize electricity prices and provided significant economic benefits to the local community.

In India, the Jaisalmer Wind Farm, located in the state of Rajasthan, has an installed capacity of approximately 1.6 GW. This project is located in a desert region and takes advantage of strong winds to generate electricity. The Jaisalmer wind farm has played a crucial role in expanding India's renewable energy capacity and reducing its dependence on fossil fuels. In addition, it has generated employment and development in a rural region, improving the quality of life of its inhabitants.

Advances in the design and efficiency of onshore wind turbines have significantly improved their viability and attractiveness as a renewable energy source. Successful projects around the world demonstrate the potential of onshore wind to contribute to the reduction of carbon emissions, improve energy security and provide economic and social benefits to communities. With continued innovation and supportive policies, onshore wind turbines will continue to play a crucial role in the transition to a sustainable energy future.

Offshore Wind Energy

Offshore wind has emerged as a powerful solution for clean and sustainable electricity generation. Located in areas with constant strong winds, offshore wind installations can take advantage of more abundant and reliable wind resources than their onshore counterparts. Large-scale turbines, typical of offshore projects, are generally larger and more powerful than onshore turbines, with capacities exceeding 10 MW per unit. Recent models, such as General Electric's Haliade-X, can reach up to 14 MW. These turbines are designed to withstand harsh environmental conditions, such as high winds, saltwater corrosion, and heavy waves, using advanced materials, including composites and specialty steels, to ensure durability and reliability.

Marine turbine foundations and supports have also advanced significantly. The most common support structures include monopiles, large steel columns driven into the seabed, and jackets, steel lattice structures fixed to the seafloor. These structures provide stability and support for turbines in relatively shallow water. However, in deep waters, where fixed structures are not viable, floating platforms have emerged as an innovative solution. These platforms anchored to the seabed allow the installation of turbines in waters more than 60 meters deep, significantly expanding the location potential of offshore wind farms.

Electricity transmission from offshore wind farms to the coast is done through submarine cables designed to handle large transmission capacities and withstand underwater conditions. Offshore substations concentrate and transform the electricity generated before sending it ashore, optimizing transmission efficiency and reducing energy losses. The installation and maintenance technology of these facilities has also advanced, using specialized vessels capable of transporting and assembling giant components on the high seas. These vessels are equipped with high-capacity cranes and dynamic positioning systems. In addition, underwater drones and robots are revolutionizing inspections and maintenance, enabling real-time monitoring and task performance without the need for direct human intervention.

Offshore wind faces several unique challenges due to its operating environment. Offshore turbines must withstand high winds, high waves, and saltwater corrosion, factors that can accelerate wear and damage to equipment. To overcome these challenges, corrosion-resistant materials and protective coatings are used to improve the durability of the turbines. In addition, robust structural designs and advanced anchoring techniques ensure stability in harsh conditions. The installation and maintenance costs of offshore wind farms are significantly higher than those of onshore installations due to the maritime conditions and logistics involved. However, economies of scale and technological innovations are driving down these costs. The adoption of specialized installation vessels, drones for inspection, and underwater robots for maintenance is improving efficiency and reducing operating expenses.

The construction and operation of offshore wind farms can affect marine life, birds, and fishing activities, and may also face opposition from local communities and maritime industries. To minimize these effects, detailed environmental impact studies are carried out and mitigation measures are implemented. Collaboration with local communities and stakeholders, as well as careful design of wind farms to avoid sensitive areas, are crucial for social acceptance. Transmitting electricity from offshore wind farms to the onshore grid requires undersea cable infrastructure and offshore substations, which can be complex and expensive. However, innovation in high-capacity cable technologies and floating substations is improving transmission efficiency. Advances in network integration and strategic infrastructure planning are also helping to overcome these challenges.

Wind variability can affect the consistency of offshore wind power generation, posing challenges to grid stability. Integrating energy storage systems, such as high-capacity batteries, and implementing smart grid management systems help balance supply and demand. In addition, offshore wind farms are often located in areas with more consistent and predictable winds, which improves reliability.

The Hornsea One Wind Farm in the UK is a leading example of success in offshore wind. Located in the North Sea off the coast of Yorkshire, Hornsea One is the world's largest offshore wind farm, with an installed capacity of 1.2 GW. This farm provides electricity to more than one million homes in the UK and has set new standards for the scale and efficiency of offshore wind. In the United States, the Block Island Offshore Wind Farm, located off the coast of Rhode Island, is the nation's first commercial offshore wind farm, with a capacity of 30 MW. This project has demonstrated the viability of offshore wind in the United States and paved the way for future developments in the region.

Another innovative project is the Hywind Scotland Offshore Wind Farm in the United Kingdom. This is the world's first floating wind farm, with an installed capacity of 30 MW, located off the coast of Scotland. Hywind Scotland has demonstrated the potential of floating platform technologies, enabling the installation of turbines in deep waters and expanding the geographical reach of offshore wind. The Walney Extension Offshore Wind Farm, also in the UK, with an installed capacity of 659 MW, is one of the largest offshore wind farms in the world, located in the Irish Sea. This project has contributed significantly to the supply of clean energy in the UK and has demonstrated the ability of large offshore wind installations to generate electricity at scale.

These success stories show how offshore wind is transforming electricity generation and setting new standards in the industry. With continuous innovation and the development of new technologies, offshore wind has the potential to play a crucial role in the transition to cleaner and more sustainable energy, taking advantage of the vast wind resources available in the world's seas.

Emerging Technologies

Emerging technologies in the field of wind energy are opening up new frontiers and opportunities, especially in areas where traditional solutions face limitations. Among these innovations, floating wind turbines stand out as one of the most promising, but there are also other technologies in development that could transform the wind energy landscape. Floating wind turbines are designed to be installed in deep waters, where traditional fixed-bottom wind turbines are not viable. These turbines are mounted on floating platforms that are anchored to the seabed with cables and ballasts. There are several floating platform designs, including semi-submersible systems, anchored tension platforms (TLPs), and spar platforms. Each design has its own advantages in terms of stability, costs, and ease of installation. Several pilot projects and prototypes have been developed to demonstrate the feasibility of floating wind turbines. For example, Hywind Scotland, operated by Equinor, is the world's first commercial floating wind farm and has shown promising results in terms of efficiency and stability.

Vertical axis turbines (VAWTs) feature an alternative design to horizontal axis turbines (HAWTs). Instead of having blades that rotate around a horizontal axis, VAWTs have blades that rotate around a vertical axis. VAWTs can capture wind from any direction, making them especially useful in areas with turbulent or changing winds. In addition, its compact design and lower center of gravity can reduce installation and maintenance costs. These turbines are being developed for urban applications, where space is limited and wind patterns are more chaotic, as well as for remote and offshore locations.

Another promising innovation is high-altitude wind turbines, designed to take advantage of the stronger and more constant winds found at higher altitudes. They use balloons, kites, or drones to lift generators hundreds of feet above the ground. Companies such as Makani, acquired by Google, have been developing kites equipped with generators that capture wind energy at high altitude. These technologies are in experimental phases but promise greater efficiency and generation capacity compared to conventional onshore turbines.

Modular and detachable turbines are designed to be easily transportable and installable in various locations. These turbines can be assembled and dismantled quickly, making them ideal for temporary projects or in hard-to-reach areas. These technologies are especially useful for emergency operations, rural development projects, and situations where temporary but reliable power is required. The flexibility and portability of these turbines allow for rapid deployment, providing an effective solution to meet energy needs in changing situations.

Emerging technologies in wind energy present great potential to transform renewable energy generation and significantly expand the reach and capacity of wind energy. Floating wind turbines allow the exploitation of wind resources in deep waters, where the winds are stronger and more constant. This could unlock vast areas of the ocean for wind power generation, reducing competition for space on land and in coastal waters. Vertical axis turbines and other compact technologies can be installed in urban environments and in remote communities, providing a local, decentralized source of renewable energy.

The aerodynamic optimization of turbine blades and structures is increasing energy capture efficiency, reducing losses and improving the profitability of wind projects. The integration of energy storage technologies, such as advanced batteries and thermal storage systems, is improving the reliability and stability of wind generation, enabling greater penetration of wind energy into the grid. As emerging technologies mature and are deployed on a larger scale, production, installation, and maintenance costs are decreasing. Economies of scale and standardization of components will contribute to making wind energy even more competitive.

The use of new materials and advanced manufacturing techniques is reducing the costs of wind turbines, improving their durability and making them easier to transport and install. The mass adoption of emerging wind technologies will contribute significantly to the reduction of greenhouse gas emissions, supporting global decarbonization and climate change mitigation goals.

Emerging wind energy will increasingly integrate with other renewable energy sources, such as solar and storage, creating hybrid energy systems that maximize efficiency and reliability.

The expansion of emerging wind energy will generate new employment opportunities in research, development, manufacturing, installation and maintenance. Decentralized wind technologies can provide access to clean and affordable energy to remote and developing communities, improving quality of life and fostering local economic development. Continuous innovation and the implementation of pioneering projects in the field of wind energy are opening new possibilities and expanding the potential of renewable energy generation. With the advancement of these technologies, wind energy is expected to play a crucial role in the transition to a cleaner, more sustainable and equitable energy future.

Chapter 3: Emerging Renewables

Tidal Energy

Tidal energy, or tidal energy, harnesses the movement of the tides to generate electricity. This type of renewable energy is based on the gravitational pull of the Moon and Sun on Earth's oceans, causing variations in sea level known as tides. Tides are periodic movements of sea level caused by the gravitational forces of the Moon and Sun, and these variations in water level can be predictable and regular, providing a constant renewable energy source. Tidal currents, which are flows of water generated due to differences in sea level during high and low tides, can be harnessed to generate electricity by submerged turbines.

Tidal current turbines work similarly to wind turbines, but they are submerged in water and driven by tidal currents. These turbines can be horizontal or vertical axis and are designed to operate in difficult underwater conditions, generating electricity at both incoming and outgoing tide. Another technology used is tidal barriers, dam-like structures that are built across a bay or estuary. These barriers use gates and turbines to capture and convert tidal energy into electricity, controlling the flow of water for more stable and controlled electricity generation.

Tidal lagoons are man-made bodies of water created by building a barrier around a section of the sea. As the tide rises and falls, water flows in and out of the lagoon through turbines, generating electricity. These lagoons have the advantage of minimizing the environmental impact by not completely obstructing natural estuaries. Water column oscillators (OWCs) are structures that use the vertical movement of tides to generate electricity. The flow of water from the tides compresses and decompresses the air in a chamber, which spins a turbine connected to a generator.

Tidal energy has been the subject of several pilot projects and case studies around the world, demonstrating the technical and economic feasibility of this renewable energy source, although it still faces challenges for its large-scale implementation. The Rance Tidal Power Station in France, opened in 1966, is the oldest and one of the largest tidal power plants in the world. Located on the estuary of the Rance River in Brittany, France, the plant has an installed capacity of 240 MW. La Rance has demonstrated the reliability and longevity of tidal barrier technology, generating enough electricity to power approximately 130,000 homes and contributing significantly to the reduction of carbon emissions in the region.

The MeyGen project in Scotland, located in the Pentland Firth Strait, is one of the largest tidal power projects in the world, with an installed capacity of 6 MW in its initial phase and plans to expand to 398 MW. MeyGen has demonstrated the viability of tidal current turbines in challenging conditions, generating valuable data on turbine performance and contributing to the development of the tidal power industry in the UK. The Swansea Bay Tidal Lagoon project in the UK was designed to be the first tidal lagoon in the world, with a planned capacity of 320 MW. Although the project has not advanced to the construction phase due to financial and regulatory challenges, it has generated significant interest in tidal lagoons as a viable option for renewable energy generation.

The Sihwa Lake tidal power plant in South Korea, inaugurated in 2011, is currently the largest in the world, with an installed capacity of 254 MW. The plant uses a tidal barrier built on Lake Sihwa on South Korea's west coast. The Sihwa Lake plant has demonstrated the viability of large-scale tidal barriers and provided a significant source of renewable energy in South Korea, as well as helping to improve water quality in the lake by controlling water flow. The Tidal Lagoon Power project in the UK includes the construction of several tidal lagoons around the British coast, with each lagoon potentially generating between 200 and 300 MW of electricity. Although the projects are in various stages of planning and approval, they represent an innovative approach to tidal power generation, and if implemented, could contribute significantly to the UK's renewable energy targets and provide valuable lessons for future tidal lagoon installations around the world.

These pilots and case studies show how tidal energy can complement other renewable technologies, such as solar and wind, and bring significant benefits to the global energy supply. Through a combination of advanced technologies and deployment strategies, tidal energy has the potential to become an integral part of the renewable energy matrix, providing a consistent and predictable source of clean and sustainable electricity. With continued support in research and development, as well as policy and financing, tidal energy can overcome its current challenges and play an important role in the transition to a cleaner and more sustainable energy future.

Wave Energy

Wave energy is a promising source of renewable energy that harnesses the movement of ocean waves to generate electricity. This form of energy is based on the conversion of the kinetic and potential energy of waves into electricity, using a variety of technologies and devices designed to capture and transform the movement of water into usable energy. Wave energy has enormous potential due to the vast surface of the oceans and the constant energy generated by the wind blowing over the water.

There are several technologies and devices used to capture wave energy, each with its own specific characteristics and applications. One of the most common devices is the absorption point wave energy converter. These devices float on the surface of the water and move with the waves, using that movement to power generators that produce electricity. Absorption point converters can be designed in different shapes, including buoys and floating platforms, and are suitable for a wide variety of ocean conditions.

Another type of technology is the water column oscillator (OWC), which uses the vertical movement of waves to generate electricity. OWCs consist of partially submerged chambers that capture compressed and decompressed air by the movement of waves. This airflow moves a turbine connected to a generator, producing electricity. OWCs can be installed on shore, in concrete structures, or on floating offshore platforms, offering flexibility in their implementation.

Oscillating body wave energy converters are another important technology. These devices consist of multiple articulated sections that float on the surface of the water and move with the waves. The articulation between the sections generates relative motion, which is converted into electricity by hydraulic or mechanical generators.

Oscillating body converters can be designed to adapt to different ocean conditions and are effective in capturing wave energy offshore.

In addition to these technologies, attenuator wave energy converters and overpass devices also play a role in capturing wave energy. Attenuators are long, floating devices that align perpendicular to the direction of the waves. As waves pass along the attenuator, the device flexes and generates power through hydraulic or mechanical systems. Overpass devices, on the other hand, capture water from waves in an elevated structure and then allow the water to flow downward through a turbine to generate electricity. This technology resembles the operation of a hydroelectric dam but uses wave movement rather than a constant river flow.

Despite the enormous potential of wave energy, there are several challenges that need to be overcome for its large-scale implementation. One of the main challenges is the durability and reliability of devices in the marine environment. Ocean conditions are extremely harsh, with high waves, strong currents, and corrosive salt water. These factors can cause significant wear and tear on devices and require frequent and costly maintenance. In addition, wave energy devices must be able to withstand extreme conditions, such as storms and storm surges, without sustaining major damage.

Another major challenge is the cost of installing and maintaining wave energy devices. Currently, wave energy technology is more expensive compared to other renewable energy sources, such as solar and wind. The costs of manufacturing, installing, and maintaining devices can be prohibitively high, especially in the early stages of development and deployment. However, costs are expected to decrease as the technology matures and economies of scale are achieved.

In addition, the integration of wave energy into the power grid presents technical and infrastructural challenges. Generating electricity from waves can be variable and depends on sea conditions, which can make it difficult to forecast and manage energy supply. Existing transmission and distribution infrastructures may need to be upgraded or adapted to handle electricity generated in coastal and remote locations.

Despite these challenges, the outlook for wave energy is promising. Continuous research and development is leading to significant improvements in the efficiency and durability of devices. Advances in coating materials and technologies are helping to protect devices from corrosion and wear, extending their lifespan and reducing maintenance costs. In addition, efforts to standardize components and manufacturing processes are helping to reduce costs and accelerate the deployment of technology.

Public-private collaboration is also driving wave energy development. Governments in several countries are providing funding and support for research and demonstration projects, while private companies are investing in the development of innovative technologies and solutions. This collaboration is helping to overcome technical and economic challenges and promote the adoption of wave energy as a viable source of renewable energy.

In the future, wave energy has the potential to play an important role in the transition to a more sustainable energy matrix. As technology continues to advance and costs decrease, we are likely to see greater adoption of wave energy devices around the world. With its ability to provide a constant and predictable source of electricity, wave energy can complement other renewable sources and contribute to reduced carbon emissions and dependence on fossil fuels.

Other Energy Sources

In addition to the most common renewable energy sources, such as solar, wind, and tidal, there are other energy sources that are gaining attention and development due to their potential to contribute to a sustainable energy mix. These include geothermal, biomass, and other innovative sources that offer unique advantages and opportunities to diversify renewable energy generation.

Geothermal energy

Geothermal energy is a renewable energy source that harnesses the Earth's internal heat, from the radioactive decay of minerals and waste heat from the planet's formation process. This heat is stored at different depths and can be used both to generate electricity and for direct heat applications. The operation of geothermal systems is based on extracting this heat from the subsoil and converting it into usable energy.

The Earth's heat can be captured and converted into energy through various types of geothermal generation systems. Dry steam systems are one of the oldest and most efficient methods, using steam directly extracted from geothermal reservoirs to drive electricity-generating turbines. These systems are relatively simple and efficient, as the steam is at high pressure and temperature, allowing for a direct conversion of thermal energy into mechanical energy and then into electricity.

Another type of system is flash steam, which converts hot water extracted from the underground into steam. This method is particularly useful in regions where groundwater is at extremely high temperatures. The hot water is depressurized in a controlled manner to convert a part into steam, which is then used to drive the generating turbines. Flash steam systems are flexible and can adapt to various geothermal conditions, making the most of the available resources.

Binary cycle systems represent a newer and more advanced technology in geothermal power generation. In these systems, a secondary fluid with a lower boiling point than water is used, such as isobutane or isopentane. This fluid is heated by geothermal heat and vaporized, generating a steam cycle that drives the generating turbines. Binary cycle systems are especially useful for the use of low and medium temperature geothermal resources, expanding the possibilities of using this renewable energy.

Geothermal energy has a wide range of applications, both in electricity generation and in direct heat supply. Geothermal plants can provide a constant and reliable source of electricity, contributing significantly to the stability of the power grid. A prominent example is The Geysers geothermal plant in California, the largest geothermal complex in the world, with an installed capacity of more than 1.5 GW. This plant uses dry steam to generate electricity, powering thousands of homes and reducing dependence on fossil fuels.

Geothermal energy is also used in direct heating applications. Such applications are common in regions with abundant geothermal resources, such as Iceland. In the city of Reykjavik, most of the domestic heating is supplied by geothermal energy. The heat extracted from the underground is used to heat water which is then distributed through central heating systems, providing an efficient and sustainable solution for district heating.

Low-temperature geothermal systems have specific applications in heating and cooling using geothermal heat pumps. These systems are suitable for residential and commercial buildings and provide an efficient and sustainable solution for temperature management. Geothermal heat pumps take advantage of the relatively constant temperature of the subsurface to transfer heat into buildings in winter and extract it in summer. This allows for efficient regulation of the indoor temperature, reducing energy consumption and carbon emissions.

In addition to its application in electricity generation and direct heating, geothermal energy has great potential in other industrial sectors. For example, geothermal heat can be used in industrial processes that require high temperatures, such as drying agricultural products, heating greenhouses, and various applications in the food and beverage industry. These applications not only increase the energy efficiency of industrial processes, but also contribute to environmental sustainability by reducing reliance on non-renewable energy sources.

The development of geothermal projects also has a positive impact on local economies, creating jobs and stimulating economic development. The construction and operation of geothermal plants require a variety of skills and services, from engineering and construction to maintenance and operational management. This creates employment opportunities in local communities and fosters regional economic growth.

Geothermal energy, with its ability to provide a constant and reliable source of energy, has the potential to play a crucial role in the transition to a more sustainable energy system. As technology advances and new applications are developed, we are likely to see an expansion in the use of geothermal energy around the world, harnessing the Earth's internal heat to meet a variety of energy needs and contribute to the reduction of carbon emissions.

Biomass

Biomass is a renewable energy source that harnesses organic material, including agricultural, forestry and urban waste, as well as specific energy crops, to generate energy. This organic material, known as biomass, can be converted into energy through various processes that take advantage of its chemical and physical properties. These processes include direct combustion, gasification, pyrolysis and anaerobic digestion, each with its own advantages and specific applications.

Direct combustion is one of the most common methods of converting biomass into energy. This process involves burning organic material to produce heat, which can be used to generate steam and drive electricity-generating turbines. Biomass plants, such as the Drax plant in the UK, use wood pellets and other organic waste to generate electricity on a large scale. The Drax plant is one of the largest biomass plants in the world and has demonstrated the feasibility of this method to produce electricity sustainably.

Another conversion method is gasification, which transforms biomass into a combustible gas by applying high temperatures in a controlled environment with a limited amount of oxygen. The gas produced, known as syngas, can be used to generate electricity, heat, or as a feedstock to produce biofuels and chemicals. Gasification is particularly useful for converting wood waste and other lignocellulosic materials into usable energy, providing a clean and efficient alternative to direct combustion.

Pyrolysis is a process that breaks down biomass by heating in the absence of oxygen, producing a gas, a liquid (biooil) and a solid waste (biochar). Biooil can be refined and used as a liquid fuel, while biochar can be used as a soil amendment to improve its fertility and water-holding capacity. Pyrolysis offers a versatile way to harness different types of biomasses, especially those with high moisture content, such as agricultural residues and sewage sludge.

Anaerobic digestion is a biological process that converts organic waste into biogas through the action of microorganisms in the absence of oxygen. Biogas, composed mainly of methane and carbon dioxide, can be used to generate electricity, heat or be purified to become biomethane, a substitute for natural gas. A prominent example is the Ludlow biogas plant in the UK, which converts food waste into renewable energy and fertilizer, demonstrating an efficient way to manage organic waste and produce clean energy.

Biomass can also be used to produce liquid biofuels, such as bioethanol and biodiesel, from energy crops such as sugarcane, corn, and soybeans. Bioethanol is produced by fermenting sugars present in crops, while biodiesel is obtained from vegetable oils and animal fats through a process called transesterification. These biofuels can be used in vehicles and machinery, providing a renewable alternative to fossil fuels and helping to reduce greenhouse gas emissions.

Generating electricity from biomass is an important application that has gained popularity in recent decades. Biomass plants not only help reduce the amount of organic waste that ends up in landfills, but they also provide a reliable and consistent source of energy. In addition, biomass plants can operate in synergy with other renewable energy sources, such as solar and wind, to ensure a stable and diversified energy supply.

The production of biogas through anaerobic digestion is another key application of biomass, especially in the management of organic waste. This process not only produces renewable energy, but also generates digestate, a by-product that can be used as organic fertilizer, closing the nutrient cycle and contributing to agricultural sustainability. Biogas plants are especially beneficial in rural and agricultural areas, where organic waste is abundant and the need for energy and fertilizer is high.

In the field of liquid biofuels, the development of advanced technologies to produce bioethanol and biodiesel has opened up new opportunities for energy diversification. Biofuels can be integrated into existing transport infrastructures, reducing dependence on fossil fuels and decreasing carbon emissions. In addition, advances in biotechnology are enabling the use of non-edible raw materials, such as agricultural residues and algae, to produce biofuels, improving the sustainability and efficiency of these processes.

An outstanding initiative in the production of 100% renewable fuels is the one developed by Repsol at its plant in Escombreras, Spain. This initiative represents a significant step towards sustainability and the reduction of carbon emissions. Repsol has implemented advanced technologies to produce second-generation biofuels from organic waste and biomass. These biofuels are capable of completely replacing traditional fossil fuels in transportation and other energy applications.

The Escombreras plant focuses on the production of bioethanol and biodiesel from sustainable feedstocks, such as used vegetable oils, animal fats and agricultural waste. These renewable biofuels not only help reduce greenhouse gas emissions, but also contribute to the circular economy by valorizing waste that would otherwise be discarded. The

Escombreras plant is an example of how innovation and technology can transform energy production and make it more sustainable.

Biomass, as a renewable energy source, has great potential to contribute to the reduction of carbon emissions and the development of a circular economy. As conversion technologies advance and production costs decrease, biomass is expected to play an increasingly important role in the transition to a more sustainable energy system. With its ability to harness a wide range of organic materials and its multiple applications in electricity generation, biogas production and biofuel processing, biomass is positioned as a viable and effective option to address the energy and environmental challenges of the future.

Other Innovative Sources

Wave energy is a renewable source that harnesses the constant movement of ocean waves to generate electricity. This process uses devices that can be both floating and fixed, and operate through various mechanisms, such as oscillating motion, displacement, and pressure. These devices convert the kinetic energy of waves into electrical energy using systems that may include oscillating buoys, oscillating water columns, and other innovative designs. A historical example of this type of technology was the Pelamis wave energy facility in Scotland. Although no longer in operation, Pelamis' technology has had a significant influence on the development of new wave energy devices. Another notable project is the Wave Hub in Cornwall, UK, which provides a testing infrastructure for wave energy technology, facilitating research and development of more efficient and commercially viable solutions.

Osmotic energy, also known as blue energy, is based on the difference in salinity between seawater and freshwater found at river mouths. This salinity gradient creates an osmotic pressure difference that can be used to generate electricity through processes such as reverse osmosis. The osmotic power plant in Tofte, Norway, is a prominent example of this technology in experimental stages. Although still in an early stage of development, the plant has demonstrated the potential of osmotic energy to contribute to electricity generation. Research in this field continues, with the aim of improving the efficiency and commercial viability of the technology, exploring new ways to optimize the generation process and reduce the associated costs.

Concentrating solar power (CSP) is another innovative source of renewable energy that uses mirrors or lenses to concentrate a large area of sunlight onto a small receiver, where the solar energy is converted into heat. This heat is used to generate steam, which in turn drives a turbine to produce electricity. One of the advantages of CSP technology is its ability to include thermal storage systems, allowing electricity generation even after sunset. This makes CSP power a viable and reliable option for providing consistent electricity. A prominent example of a CSP project is the Noor plant in Morocco, one of the largest facilities of its kind in the world. The Noor plant has demonstrated how concentrating solar power can provide a renewable and reliable source of energy on a large scale, helping to reduce dependence on fossil fuels and contributing to energy sustainability.

These innovative sources of energy, although in different stages of development and deployment, represent important advances in the diversification and sustainability of the global energy matrix. Wave energy, osmotic energy and concentrating solar energy not only offer new ways to generate clean electricity, but also expand the possibilities for harnessing the natural resources available in different geographical environments. Continuous research and development in these fields are essential to improve efficiency, reduce costs and overcome the technical challenges that still exist, facilitating a faster transition to a sustainable energy future.

Potential and Future Prospects

Emerging renewable energy sources have enormous potential to complement more established technologies and contribute to a more diversified and sustainable energy mix. Geothermal energy is one of these sources with great prospects for global expansion. Geothermal resources are available in many regions of the world, especially in areas with high geothermal activity such as the Pacific Ring of Fire, which encompasses countries such as Japan, Indonesia, the Philippines, and New Zealand. These resources enable the generation of electricity and the supply of heat in a constant and reliable manner. In addition, technological innovations are driving access to deeper geothermal resources. Advances in drilling and exploration techniques are making it possible to implement enhanced geothermal energy (EGS) systems, which allow geothermal resources to be harnessed in areas where it was not previously viable.

Biomass, on the other hand, offers a sustainable way to manage organic waste and convert it into useful energy. This process not only reduces the amount of waste that ends up in landfills, but also provides a renewable energy source that can be used to generate electricity, heat, and biofuels. Improving the efficiency of conversion processes, such as direct combustion, gasification, pyrolysis and anaerobic digestion, is increasing the economic and environmental viability of biomass. Optimizing the production of energy crops, such as corn and sugarcane, is also contributing to this efficiency gain. In addition, biomass plays a crucial role in the circular economy, closing nutrient and energy cycles by converting waste into valuable resources, such as fertilizers and biofuels, thus promoting more sustainable resource management.

The other innovative sources of energy, such as wave energy and osmotic energy, are moving from the experimental phase to commercialization. Wave energy, which uses the constant movement of ocean waves to generate electricity, is being tested in pilot projects such as the Wave Hub in Cornwall, UK, and the Pelamis facility in Scotland. These projects are demonstrating the viability of the technology and attracting investment for its large-scale development. Osmotic energy, also known as blue energy, which is based on the difference in salinity between seawater and freshwater, is in an experimental stage, but the osmotic power plant in Tofte, Norway, has shown its potential. Continued research and successful pilot projects are essential to improve the efficiency of these technologies and make them commercially viable.

Integrating these innovative sources with other renewable technologies can improve the stability and reliability of the energy supply. The combination of different renewable energy sources makes it possible to create a more robust and resilient energy system, capable of adapting to variations in the availability of energy resources. For example, solar and wind power can be complemented by geothermal and biomass power, which offer more consistent power generation. In addition, wave energy and osmotic energy can be integrated into hybrid systems that optimize the use of available resources.

Diversifying renewable energy sources with emerging technologies such as geothermal, biomass, and other innovations is crucial to the transition to a sustainable energy future. Each of these sources offers unique advantages and can complement more established technologies, helping to reduce carbon emissions, improve energy security and promote sustainable economic development. Continued support for research and technological development, as well as the implementation of pilot projects, are essential to demonstrate the viability of these technologies and attract investment. With these efforts, emerging energy sources have the potential to play an increasingly important role in the global energy mix, contributing to a cleaner, more efficient and resilient energy system.

Chapter 4: Innovations in Storage and Management

Advanced Batteries

Advanced batteries are playing a crucial role in the transition to a sustainable energy future by providing energy storage solutions that improve the reliability and efficiency of renewable energy sources. Among the most prominent types are lithium-ion batteries, which are known for their high energy density, long lifespan, and charging and discharging efficiency. These batteries are widely used in energy storage applications, such as electric vehicles and backup systems. Recent advances in electrode and electrolyte chemistry have improved the capacity, safety, and longevity of lithium-ion batteries, with innovations such as silicon anodes and solid electrolytes increasing energy density and reducing the risk of fires.

Solid-state batteries, on the other hand, use solid electrolytes instead of liquids, which increases safety and energy density. These batteries promise greater thermal and chemical stability. Current research is focused on finding suitable materials for solid electrolytes that are stable and conductive, while improvements in the interface between the electrolyte and electrodes are reducing internal resistance and increasing efficiency. Redox flow batteries, which store energy in electrolyte solutions contained in external tanks, are also seeing significant developments. The energy is generated when electrolytes flow through an electrochemical cell that converts chemical energy into electricity. Recent developments in electrolyte chemistry and cell designs are increasing the efficiency and capacity of these batteries, with research focused on materials such as vanadium and zinc-bromine, which offer higher energy densities and lower costs.

Sodium-ion batteries, which use sodium instead of lithium, present a more affordable and sustainable alternative due to the abundance of sodium. Although they have a lower energy density compared to lithium batteries, their low cost and sustainability make them attractive. Advances in electrode and electrolyte materials have improved the capacity and efficiency of sodium-ion batteries, with recent research addressing challenges related to long-term stability and cyclability. On the other hand, lithium-phosphor ion (LFP) batteries are known for their high thermal stability and safety, having a longer lifespan and lower degradation compared to other lithium chemistries. Optimization of manufacturing processes and advances in electrode materials are increasing the energy density and reducing the costs of LFP batteries, making them more competitive for large-scale storage applications.

Advanced batteries are being deployed in a variety of projects and applications that demonstrate their ability to improve the efficiency and reliability of renewable energy supply. For example, the Hornsdale Power Reserve in Australia, known as the "Tesla battery of South Australia," uses lithium-ion batteries with a capacity of 150 MW/194 MWh. This installation has significantly improved the stability and reliability of the electricity grid in South Australia, reducing power outages and frequency regulation costs. In addition, it has demonstrated the viability of lithium-ion batteries for large-scale storage applications.

In China, one of the world's largest redox flow battery facilities is being developed in Dalian, with a planned capacity of 200 MW/800 MWh using vanadium flow batteries. This project aims to provide large-scale energy storage to support the integration of renewable energy sources, improve grid stability, and reduce carbon emissions. The storage capacity will allow for better management of variability in renewable energy generation.

Japan is at the forefront of solid-state battery development, with companies such as Toyota and Panasonic leading research and development for applications in electric vehicles and stationary storage. These batteries promise to revolutionize the energy storage industry with their increased energy density and improved safety. Pilots and demonstrations are laying the groundwork for large-scale commercialization and adoption in various applications.

In the United States, the Moss Landing Energy Storage Facility project in California, operated by Vistra Energy, is one of the largest energy storage facilities in the world, with a capacity of 400 MW/1,600 MWh using lithium-ion batteries. This project provides significant energy storage capacity to support California's electric grid, facilitating the integration of solar and wind energy. Installation also improves grid reliability and helps avoid power outages during peak demand.

In Europe, specifically in Spain, the GRESB project has developed an energy storage facility using lithium-phosphorus ion (LFP) batteries with a capacity of 100 MW/200 MWh. This project improves grid stability and renewable energy integration, reducing dependence on fossil fuels and decreasing carbon emissions. The choice of LFP batteries provides a safe and long-lasting solution for large-scale energy storage.

Thermal and Mechanical Storage

Energy storage is essential to manage the intermittency and variability of renewable energy sources such as solar and wind. Among the various storage technologies, thermal and mechanical solutions offer complementary and effective methods for storing energy on a large scale. One of the most prominent technologies in this field is compressed air energy storage (CAES), which works by storing energy by compressing air in underground cavities or high-pressure tanks. During periods of high demand, compressed air is released, heated, and expanded through a turbine to generate electricity. Key components of this system include compressors that use electricity to compress air, storage cavities that are typically natural geological formations such as salt caverns capable of withstanding high pressure, and expansion turbines that convert energy from compressed air into electricity. Among the best-known projects are the Huntorf CAES plant in Germany, operational since 1978 with a capacity of 290 MW, and the McIntosh CAES plant in Alabama, USA, with a capacity of 110 MW.

Another prominent technology is flywheels, which store kinetic energy in a rotor that rotates at high speed inside a low-friction container. Energy is stored when the rotor accelerates during periods of low demand and is released when it decelerates during periods of high demand. Key components of flywheels include the rotor, usually made of composite materials to minimize weight and maximize strength; the low-friction container, which often uses a vacuum to reduce air resistance; and the generator/motor, which acts both to accelerate the rotor and store energy and to decelerate it and release energy. A notable example of this technology is Beacon Power's facility in New York, USA, with a capacity of 20 MW, which provides frequency regulation services to the power grid.

Each energy storage technology has its own advantages and limitations, which should be considered when selecting the right solution for a specific application. Compressed air energy storage (CAES) offers high storage capacity, making it suitable for large-scale applications, and a long storage duration, ideal for balancing the seasonal variability of renewable sources. In addition, CAES systems have a long service life with a low degradation rate. However, they require suitable geological cavities, which limits the possible locations for their installation, and the energy conversion efficiency of traditional CAES systems is relatively low, although more modern versions, such as adiabatic CAES, can improve this efficiency. Also, building infrastructure for CAES can be expensive, especially if artificial storage cavities are required.

On the other hand, flywheels have a high-power conversion efficiency, around 85-90%, and can respond almost instantaneously to power demands, making them ideal for grid stabilization and frequency regulation applications. In addition, they have a long service life and require little maintenance due to the absence of moving parts that wear out quickly. However, the energy storage capacity of flywheels is relatively low, making them more suitable for short-term applications, and although operating costs are low, the initial investment for flywheels can be significant. Also, the need to maintain a low-friction environment and vibration sensitivity can limit installation locations.

The choice between compressed air energy storage (CAES) and flywheels depends on the specific needs of the application. For applications that require large-scale, long-duration storage, such as supporting the power grid with high penetration of renewables, CAES may be a more suitable option. On the other hand, for applications that require rapid response and grid stabilization, such as frequency regulation and management of short-term demand peaks, flywheels are an effective solution. Both technologies play a crucial role in the transition to a more sustainable and resilient energy system, complementing other energy storage solutions such as advanced batteries and thermal storage. With continuous development and technological innovation, these mechanical and thermal storage solutions will continue to improve in terms of efficiency, capacity and economic viability, contributing significantly to the stability and reliability of renewable energy-based power grids.

Electricity Network Management

Grid management and smart solutions are crucial to ensure the stability and efficiency of the electricity system, especially with the increasing incorporation of renewable energy sources. Demand management systems seek to balance electricity supply and demand by adjusting end-user consumption in response to market or grid signals. Common techniques include direct load control, where utilities can manage high-consumption devices such as heating and cooling systems in homes and businesses to reduce demand at peak times. In addition, dynamic pricing offers variable rates based on time of day or grid demand, incentivizing consumers to reduce their consumption during peak demand and increase it during periods of low demand. Demand response programs also play a crucial role, incentivizing consumers to reduce their consumption in response to signals from the grid operator, such as high prices or balancing needs.

Smart solutions, such as smart grids, integrate advanced information and communication technologies to improve the efficiency, reliability and sustainability of the electricity supply. These networks include smart meters, sensors, and real-time communication systems that enable network monitoring and control, improving responsiveness and reducing energy losses. Smart meters provide detailed data on electricity consumption in real time, making it easier to implement dynamic tariffs and allowing consumers to better manage their consumption. Energy management systems monitor, control, and optimize the performance of electrical systems, including renewable energy sources and energy storage, managing load, optimizing energy use, and integrating storage and distributed generation.

The integration of renewables into the electricity grid presents challenges due to its intermittent and variable nature. However, various technologies and strategies are enabling more effective and reliable integration of these sources. Energy storage systems are essential for mitigating the intermittency of renewables, storing excess energy generated during periods of high production and releasing it during periods of low production. Common storage technologies include advanced batteries, compressed air energy storage (CAES), and thermal storage. Smart grids and microgrids improve the grid's ability to integrate renewable energy sources through real-time monitoring and control and distributed generation management. These grids have the ability to isolate themselves from the rest of the grid in the event of failures and integrate multiple energy sources, improving the efficiency of local demand.

Energy forecasting and management systems use weather data and advanced algorithms to predict renewable energy generation more accurately, improving grid planning and operation. Active demand management adjusts electricity consumption in real-time to balance supply and demand on the grid, using tools such as dynamic pricing, demand response programs, and direct load control. Smart inverters convert direct current generated by renewable energy sources into grid-compatible alternating current, providing ancillary services such as frequency and voltage control and the ability to operate in island mode.

Supportive policies and regulatory frameworks are essential to facilitate the integration of renewables into the grid. Examples of policies include tax incentives, feed-in tariffs, renewable energy portfolio standards, and renewable energy auction programs. These policies help reduce economic and technical barriers to the integration of renewables and encourage investment in smart grid infrastructure and energy storage.

Managing the electricity grid with the integration of renewable energy offers several benefits, including reducing greenhouse gas emissions, decreasing dependence on fossil fuels, increasing grid flexibility, and improving resilience to failures and natural disasters. It also improves the operational efficiency of the network, optimizing the use of energy resources. However, there are limitations, such as the high upfront costs of deploying smart grids, storage systems, and advanced energy conversion technologies. The interoperability of multiple technologies and energy sources in a coherent and efficient grid can be complex, and the lack of appropriate policies and regulations can hinder the deployment and adoption of these advanced technologies. In addition, smart grids and IT-based energy management systems are vulnerable to cyberattacks, requiring robust security and data protection measures.

Chapter 5: Political and Economic Framework

Government Policies

Government policies play a crucial role in the promotion and adoption of renewable energy. Over the years, several countries have implemented successful policies that have boosted the development and integration of renewables into their energy matrices. In Germany, the Energiewende seeks the transition to an energy system based on renewable energies, energy efficiency and the phasing out of nuclear energy and fossil fuels. Key policies include feed-in tariffs (FITs), subsidies for the installation of renewable systems, and ambitious CO_2 emission reduction targets. Germany has achieved rapid expansion of installed renewable energy capacity, especially in solar and wind power. By 2020, more than 40% of electricity in Germany was generated from renewable sources, making the country a global leader in renewable energy and fostering technological innovation and job growth in the sector.

In China, the government has implemented several policies in its five-year plans to promote renewable energy, including subsidies, tax credits, and installed capacity targets. The 13th Five-Year Plan (2016-2020) set specific targets for installed solar and wind power capacity, as well as providing subsidies and financial support for renewable energy projects. Thanks to these policies, China has become the world's largest producer of solar and wind energy. At the end of 2020, China had an installed capacity of wind power of more than 281 GW and solar power of more than 253 GW. These policies have helped to significantly reduce the costs of renewable technologies globally.

Denmark has adopted a comprehensive strategy for the energy transition, focusing on wind energy, energy efficiency and the integration of renewable energy into the electricity system. Key policies include tax incentives, feed-in tariffs and a policy to support research and development of renewable technologies. Denmark generates approximately 50% of its electricity from wind power. Investment in renewables has boosted economic growth, created jobs, and made Denmark an exporter of wind technology.

In the United States, policies to support renewable energy include investment and production tax credits, as well as state renewable energy portfolio standards (RPS) mandates. The Production Tax Credit (PTC) and the Investment Tax Credit (ITC) have been crucial to the growth of wind and solar capacity. In addition, several states have implemented their own RPS mandates that require a percentage of electricity to come from renewable sources. These policies have led to rapid growth in installed renewable energy capacity in the U.S. By 2020, approximately 20% of electricity generation in the U.S. came from renewable sources, with significant growth in wind and solar power.

India has set ambitious national targets for renewable energy capacity and has implemented a system of energy auctions to reduce costs and encourage investment. Key policies include a target of installing 175 GW of renewable capacity by 2022 and the implementation of competitive auctions for solar and wind power projects. India has seen a massive expansion in its renewable capacity, reaching 89 GW of installed renewable energy capacity by the end of 2020. Auctions have significantly reduced the cost of solar and wind power, making them competitive with fossil fuels.

The impact of government policies on the adoption of renewable energy is significant and multifaceted. Policies such as subsidies, feed-in tariffs, and competitive auctions have incentivized the installation of large volumes of renewable capacity, creating economies of scale that have significantly reduced the costs of solar and wind technologies. Government support has fostered competition and innovation in the renewable energy sector, leading to technological improvements and reduced production costs.

Clear and ambitious national targets have provided a clear direction and purpose for renewable energy development. This has attracted investment and facilitated long-term planning. Auctions and feed-in tariffs have been effective in ensuring fair and stable prices for renewable energy producers, thereby incentivizing investment in new capacity. Policies to support renewable energy have fostered the growth of a robust economic sector, creating jobs in the manufacture, installation, operation, and maintenance of renewable energy systems. In many countries, the development of renewable energy projects has boosted economic growth in rural and less-developed areas, providing income and improving local infrastructure.

Government policies have accelerated the transition from fossil fuels to renewable energy, significantly reducing greenhouse gas emissions and contributing to global climate goals. The adoption of renewable energy has had additional environmental benefits, such as reducing air and water pollution, and preserving natural ecosystems. However, the effectiveness of policies depends on their coherence and continuity. Sudden or inconsistent changes in policies can create uncertainty and discourage investment. As the penetration of renewables increases, it is crucial to develop and implement solutions for grid integration, such as energy storage, smart grids, and demand side management. It is important to ensure that the benefits of renewable energy policies are equitably distributed and that all segments of society have access to clean and affordable energy.

In short, government policies have been instrumental to the growth and adoption of renewable energy around the world. Examples of successful policies demonstrate that the right support can catalyze innovation, reduce costs, increase installed capacity, and generate significant economic and environmental benefits. As the world grapples with the challenges of climate change and the energy transition, government policies will continue to play a crucial role in promoting a sustainable and resilient energy future.

Economic Incentives

Economic incentives are key tools that governments use to encourage the adoption of renewable energy and accelerate the transition to a sustainable energy system. These incentives include grants, tax credits, feed-in tariffs, energy auctions, and other mechanisms that reduce the cost of investment and improve the profitability of renewable energy projects.

Grants are direct financial contributions from the government to cover part of the cost of installing renewable energy projects. These can be granted to companies, organizations or individuals. For example, the U.S. Department of Energy (DOE) grant program provides funding for renewable energy research and development projects, as well as for the implementation of pilot and demonstration projects. Subsidies reduce upfront costs, facilitating the adoption of clean technologies and fostering innovation in the energy sector.

Tax credits are also an effective tool to promote the adoption of renewable energy. The Production Tax Credit (PTC) provides a tax credit for each kilowatt-hour (kWh) of electricity generated by renewable sources over a given period. In the United States, the PTC has been instrumental in the growth of wind energy, offering a tax credit per kWh of electricity produced by wind projects during their first ten years of operation. The Investment Tax Credit (ITC), on the other hand, allows developers of renewable energy projects to deduct a percentage of the project's installation cost from their federal taxes. This incentive has been especially important for solar in the United States, allowing developers to deduct up to 26% of the total cost of installing solar systems through 2022, with a gradual reduction scheduled in the following years.

Feed-in Tariffs (FITs) guarantee renewable energy producers a fixed price for each kWh of electricity they inject into the grid, usually for an extended period. Germany has used feed-in tariffs since the early 2000s to boost solar and wind power. These tariffs have provided stability and predictability to investors, facilitating a rapid expansion of installed capacity. Feed-in tariffs ensure stable incomes for producers, incentivizing investment in new renewable energy capacities.

Energy auctions are competitive processes in which developers of renewable energy projects submit bids to supply electricity to the grid at a certain price. Contracts are awarded to the most competitive bids. In Brazil, energy auctions have been used to allocate long-term contracts for wind, solar and biomass power generation, resulting in competitive prices and a significant expansion of renewable capacity. Auctions encourage competition among developers, reducing generation costs and promoting efficiency in the renewable energy market.

Low-interest loans and financing provide capital to renewable energy project developers at reduced interest rates, reducing the financial cost of investments. The European Investment Bank (EIB) offers financing to renewable energy projects across Europe, supporting the implementation of clean and sustainable technologies.

These financing mechanisms lower the barrier to entry for new projects and accelerate the adoption of large-scale renewables.

The effectiveness and sustainability of economic incentives vary depending on the context and specific implementation. Economic incentives reduce financial barriers to the implementation of renewable energy projects, making them more accessible and attractive to investors. ITC in the United States has been highly effective in reducing the upfront costs of solar projects, encouraging rapid growth in installed capacity. Grants and tax credits have encouraged research and development of new technologies, accelerating innovation and continuous improvement in the renewable energy sector. DOE grants have supported the development of advanced energy storage technologies and improvements in the efficiency of solar panels. Feed-in tariffs and auction contracts provide stability and predictability for investors, making it easier to plan and finance long-term projects. Feed-in tariffs in Germany have been instrumental in the expansion of the solar sector, providing a stable framework that has attracted significant investments.

Tax incentives, such as tax credits and subsidies, come at a cost to governments. The sustainability of these incentives depends on the country's fiscal capacity and long-term economic justification. In some cases, the high fiscal costs associated with incentives have led to debates about their long-term viability and the need for adjustments. Energy auctions are considered one of the most efficient ways to allocate resources, as they encourage competition and can result in lower prices for renewable electricity. Auctions in Brazil have achieved competitive prices for wind and solar power, reducing costs for consumers and promoting sustainable growth of the sector. Economic incentives not only promote the adoption of renewable energy, but can also lead to broader economic benefits, such as job creation and industrial development. The development of the wind industry in Denmark has generated thousands of jobs and made the country a leading exporter of wind technology.

The ability of incentives to adapt to market conditions and evolve over time is crucial to their sustainability. Policy review and adjustment mechanisms can ensure that incentives remain effective and efficient. Regular reviews of feed-in tariffs in Germany have made it possible to adjust incentive levels in response to market developments and reduced technology costs. The effective and sustainable implementation of economic incentives is essential for the transition to an energy system based on renewable energies. While incentives have proven to be highly effective in reducing costs, attracting investment, and fostering innovation, it is also important to consider their fiscal sustainability and ability to adapt to changing market conditions.

In conclusion, economic incentives, such as grants, tax credits, feed-in tariffs, and energy auctions, have been crucial instruments in promoting the adoption of renewable energy around the world. Its effectiveness depends on proper design, consistent implementation, and an ability to evolve over time. By addressing sustainability and efficiency challenges, governments can continue to use these incentives to drive renewable energy growth and contribute to a sustainable and resilient energy future.

Financing Strategies

Financing renewable energy projects is crucial for their development and expansion. With the evolution of the sector, innovative business models and financing mechanisms have emerged that facilitate investment and reduce the associated risks. A prominent model is that of Power Purchase Agreements (PPAs), which are contracts between a renewable energy generator and a buyer, usually a company or utility. In these contracts, the buyer agrees to purchase the electricity generated at a fixed price for a specified period. PPAs provide revenue predictability for project developers, reduce market risks, and facilitate financing by ensuring a steady revenue stream. Companies such as Google and Amazon have signed long-term PPAs to secure the supply of renewable energy for their global operations, ensuring a steady and predictable supply of clean energy for their activities.

Crowdfunding is another innovative financing strategy, allowing small individual investors to finance renewable energy projects through online platforms. Investors receive a share of the revenue generated by the project, democratizing investment in renewable energy and facilitating community participation. Abundance Investment in the UK has financed multiple renewable energy projects through crowdfunding, allowing citizens to invest directly in clean energy and actively participate in the energy transition.

Green Bonds are debt instruments issued to finance projects with environmental benefits, such as renewable energy projects. Investors receive a fixed return on their investment, similar to traditional bonds, but with the added bonus of contributing to sustainability. These bonds attract a wide range of investors interested in sustainable investments, offering transparency on the use of funds and providing competitive interest rates. The World Bank has issued multiple green bonds to finance renewable energy and sustainability projects around the world, mobilizing large sums of capital towards green initiatives.

Renewable energy investment funds are another effective tool. These specialized funds raise capital from multiple investors to fund a diversified portfolio of clean energy projects. Risk diversification and access to large-scale projects are key advantages of this approach, along with professional investment management. Brookfield Renewable Partners, for example, manages one of the largest renewable energy investment funds, with diversified assets in wind, solar, hydropower, and other technologies, demonstrating the funds' potential to mobilize large amounts of capital toward sustainable projects.

Community-based business models are also an effective strategy. Community-based renewable energy projects are funded and often operated by local community groups. The financial and energy benefits are maintained within the community, promoting local acceptance and encouraging community participation. The Gigha Energy Co-operative in Scotland is a prominent example of a community project that operates wind turbines and reinvests the proceeds into local initiatives, creating a positive impact on both the environment and the local economy.

The Noor Solar Energy Project in Morocco is an exemplary case study. The Noor solar complex in Ouarzazate is one of the largest solar projects in the world, with a planned capacity of 580 MW, using concentrating solar technology (CSP) and solar photovoltaics. This project was financed through a combination of loans from the World Bank, the African Development Bank and the European Investment Bank, along with funds from the Moroccan government. Noor has demonstrated the potential of large solar projects in Africa, improving energy security and reducing carbon emissions in the region.

Another prominent case is the Gansu Wind Farm in China, one of the largest in the world, with an installed capacity of more than 10 GW. This project was financed through a combination of Chinese government investments, bank loans, and private equity. It has contributed significantly to China's renewable energy capacity, reducing dependence on fossil fuels and improving air quality. These projects show how the right mix of public and private financing can drive the energy transition.

In India, the Kurnool Ultra Mega Solar Project is one of the largest solar parks in the world, with an installed capacity of 1,000 MW. Financed through a combination of Indian government funds, World Bank loans, and power purchase agreements (PPAs) with local utilities, this project has significantly increased India's solar capacity, providing clean electricity to millions of homes and helping to meet the country's ambitious renewable energy targets.

The Beacon Power Flywheel in the USA is another innovative example. Beacon Power has developed flywheel energy storage facilities in New York, with a capacity of 20 MW. Financed through a combination of private equity, grants from the U.S. Department of Energy, and bank loans, this facility improves the stability of the power grid and provides frequency regulation services, demonstrating the commercial viability of flywheels as an energy storage technology.

The Roscoe Wind Farm in Texas is one of the largest onshore wind farms in the world, with a capacity of 781.5 MW. Funded through a combination of private equity, federal tax credits and long-term power purchase agreements, Roscoe has helped diversify Texas' energy mix, providing a significant source of clean energy and reducing greenhouse gas emissions.

In summary, innovative business models and financing mechanisms are essential for the development and expansion of renewable energy projects. From power purchase agreements and green bonds to community financing and specialized investment funds, these approaches have proven effective in mobilizing capital and reducing risk. Case studies of successfully funded projects highlight the importance of the right mix of public and private financing, community engagement, and government support to achieve sustainable and meaningful growth in the renewable energy sector.

Chapter 6: The Role of Artificial Intelligence and Big Data

Power System Optimization

Artificial intelligence (AI) is revolutionizing the way energy generation and consumption are predicted and managed, providing advanced tools to improve the efficiency, reliability, and sustainability of energy systems. AI, using machine learning techniques, uses historical and weather data to create accurate forecast models that predict renewable energy generation. These models can be continuously adapted and improved as more data is collected. Accurately predicting renewable energy generation allows grid operators to better plan and manage the variability inherent in solar and wind energy sources. This reduces reliance on backup power plants and improves grid stability, which is crucial for an efficient transition to cleaner energy.

AI-based energy management systems (EMS) optimize energy consumption in real-time, adjusting demand based on electricity prices, renewable energy availability, and user needs. AI can analyze consumption patterns and activate demand response programs, where consumers reduce or shift their electricity use in response to price signals or economic incentives. This intelligent demand management reduces consumption peaks, lowers energy costs, and helps balance supply and demand in the power grid, improving overall system efficiency. This is particularly important in densely populated urban areas where energy demand can vary significantly throughout the day.

AI is also applied in the optimization of energy storage. Smart storage systems manage battery usage and charging/discharging to maximize battery efficiency and lifespan. This involves deciding when to store the excess energy generated and when to release that energy to meet demand. Optimizing energy storage ensures that energy resources are used efficiently, improves the integration of intermittent renewables, and reduces the need for fossil fuel-based backup generation. Advances in battery technology, such as lithium-ion batteries and redox flow batteries, benefit significantly from AI to maximize their performance and extend their lifespan.

AI-based predictive maintenance improves the reliability of power systems, extends equipment life, and reduces operating costs. AI can continuously monitor the performance of energy equipment and systems, using predictive analytics to identify potential failures before they occur. AI algorithms can recommend proactive maintenance actions based on real-time data analysis, reducing downtime and repair costs. This is essential for large renewable energy installations where unplanned maintenance can be costly and disruptive.

In the integration of renewable energies, AI plays a crucial role in the management of smart grids. Coordinates distributed generation, energy storage, and consumption to optimize grid performance. AI can help power grids adapt quickly to changes in generation and demand, improving system flexibility and resilience. The efficient integration of renewable energies reduces greenhouse gas emissions, improves the sustainability of the energy system and facilitates the transition to a cleaner energy mix. Smart grids that use AI can better manage the variability of solar and wind power, optimizing resource use and reducing operating costs.

The AI at the California Electric Grid (CAISO) project is a prominent example of how AI can improve energy management. The California Independent System Operator uses AI algorithms to predict solar and wind power generation, as well as to manage energy demand in real time. The implementation of AI has improved the accuracy of generation forecasts, optimized resource dispatch, and reduced the need for backup power generation, contributing to a more efficient and sustainable grid. This has enabled greater integration of renewables into California's power grid, reducing emissions and improving supply stability.

Google DeepMind has collaborated with the UK's National Grid to use AI to optimize wind energy use. AI has enabled better prediction of wind power generation and more efficient management of energy flow, helping to reduce costs and improve grid stability. Google's DeepMind AI platform can analyze large volumes of weather and power generation data, providing accurate predictions that enable grid operators to more effectively manage wind energy variability.

Siemens has developed an AI-based energy management system that is used in several renewable energy and storage projects around the world. This system optimizes battery charging and discharging, manages the use of renewable energy, and provides demand response services, improving operational efficiency and reducing carbon emissions. The implementation of this system in various facilities has demonstrated significant improvements in energy management and has contributed to the greater integration of renewable energy sources into electricity grids.

In Australia, several utilities have implemented AI-based smart grid solutions to manage renewable energy integration and improve grid reliability. AI has improved the grid's ability to manage solar and wind energy variability, optimizing the use of energy resources and reducing operating costs. These solutions include the use of advanced sensors and real-time communication systems that allow for more effective and adaptive management of the electricity grid, facilitating the integration of large volumes of renewable energy.

Artificial intelligence is transforming the way power systems are optimized, providing advanced tools to predict and manage energy generation and consumption more efficiently. The implementation of AI in demand management, energy storage, predictive maintenance, and the integration of renewable energy is driving the transition to a more sustainable and resilient energy system. Case studies and real-world applications demonstrate the enormous potential of AI to improve operational efficiency, reduce costs, and increase the reliability of energy supply around the world.

Predictive Maintenance

Predictive maintenance uses Big Data and artificial intelligence (AI) to anticipate equipment and infrastructure failures, enabling organizations to perform proactive maintenance and avoid costly downtime. This methodology is revolutionizing the management of energy infrastructures, improving the efficiency, reliability and useful life of systems. Advanced sensors installed in equipment and infrastructure collect real-time data on various operating variables, such as temperature, vibration, pressure, humidity, and performance. The large volumes of data collected are analyzed using Big Data algorithms and machine learning techniques to identify patterns and anomalies that may indicate a possible failure. AI creates predictive models that can predict when a failure is likely to occur, based on historical data and current equipment conditions.

Monitoring vibrations in rotating equipment, such as turbines and motors, can identify signs of wear or misalignment. AI algorithms analyze vibration data to detect anomalies and predict failures, enabling proactive maintenance and reducing the risk of further damage. Temperature monitoring in electrical and mechanical equipment helps identify overheating issues, which can be indicative of impending failures. AI thermal data analysis makes it possible to detect overheating patterns and predict failures before they occur, improving the reliability and safety of systems. Analysis of the chemical and physical properties of oils and lubricants can reveal the presence of contaminants and component wear. AI models can predict the need for oil changes and component maintenance based on the analysis of lubrication data, optimizing maintenance cycles and reducing costs. Corrosion sensors monitor the condition of metal surfaces in infrastructures such as pipelines and marine structures. AI analyzes sensor data to identify early signs of corrosion and predict its progression, enabling timely interventions and preventing structural failures.

Predictive maintenance allows problems to be identified and fixed before they cause failures, significantly reducing unplanned downtime. In a wind power plant, predictive maintenance can detect turbine anomalies, allowing for scheduled repairs that prevent service interruption. By performing maintenance only, when necessary, organizations can reduce the costs associated with excessive preventative maintenance and emergency repairs. A utility can optimize the maintenance cycles of its transmission and distribution equipment, reducing operating costs and improving efficiency. Early detection of problems allows interventions that can extend the life of equipment and improve its performance. In a solar installation, continuous monitoring and predictive analytics can identify problems with inverters, enabling repairs that extend the life of these critical components. The ability to predict failures improves the reliability of infrastructures and reduces the risk of accidents and catastrophic failures. In the petrochemical industry, predictive maintenance can identify valve and piping problems before they cause leaks or explosions, improving operational safety.

General Electric (GE) uses its Predix industrial data analytics platform to monitor and predict the performance of its wind turbines, aircraft engines, and other industrial equipment. The Predix platform has enabled GE to improve the reliability and efficiency of its equipment, reducing maintenance costs and improving operational performance. Siemens uses its MindSphere industrial IoT platform to collect and analyze data from its energy and automation systems. MindSphere enables Siemens and its customers to implement predictive maintenance strategies, optimizing the operation of critical infrastructures and reducing downtime. IBM Watson uses AI and big data analytics to monitor and predict the performance of solar and wind farms. IBM Watson technology has improved the operational efficiency of renewable energy projects, enabling proactive interventions and reducing maintenance costs. Royal Dutch Shell uses AI techniques and Big Data analytics to monitor its oil and gas production and distribution infrastructures. The implementation of predictive maintenance has enabled Shell to improve the reliability of its operations, reduce downtime and optimize operating costs.

Predictive maintenance is transforming the management of energy infrastructures through the application of Big Data and artificial intelligence. By anticipating failures and optimizing maintenance interventions, these technologies improve operational efficiency, reduce costs, and extend equipment life. Case studies and real-world applications demonstrate the enormous potential of predictive maintenance to improve the reliability and sustainability of energy systems around the world. This transformation is leading to greater adoption of advanced technologies in various sectors, contributing to a more efficient and sustainable future.

Data Analysis and Decision Making

The analysis of large volumes of data is crucial to optimize the management of renewable energy projects. Using advanced tools and analytical methods, organizations can extract valuable insights to improve operational efficiency, predict trends, and make informed decisions. Big Data platforms such as Apache Hadoop enable distributed processing of large data sets across clusters of computers, while Apache Spark provides an easy-to-use programming interface and is known for its speed and in-memory analysis capability. NoSQL databases such as MongoDB, which stores data in JSON document format, and Cassandra, which offers high availability and scalability, are essential for managing unstructured data. Analytics and visualization tools like Tableau and Power BI make it easy to create interactive dashboards to explore and understand large data sets, while machine learning and AI tools like TensorFlow and Scikit-learn make it easy to build and train machine learning models.

Data analytics can take a variety of methods, from descriptive analytics, which provides an overview of historical data to understand what has happened in the past, to predictive analytics, which uses statistical models and machine learning algorithms to forecast future trends and likely events based on historical data. In addition, prescriptive analytics suggests specific actions that can be taken to reach a desired outcome, based on data analytics and predictive modeling, while real-time analytics processes and analyzes data as it is generated to provide immediate insights and facilitate real-time decision-making.

The implementation of these tools and methods in renewable energy projects is diverse. For example, wind farm operators use data analytics to optimize power generation and reduce operating costs. At the Horns Rev Wind Farm in Denmark, data analysis has made it possible to improve operational efficiency and increase energy production by 5%. Solar plants use data analytics to monitor the performance of solar panels and predict maintenance issues. The Noor solar plant in Morocco uses real-time data analytics to monitor and optimize the performance of its concentrating solar systems, improving the overall efficiency of the system.

Utilities also use data analytics to manage the integration of renewables into the power grid and maintain grid stability. In California, the Electric Grid Operator (CAISO) uses AI models to forecast solar and wind power generation, improving grid management and reducing the need for backup generation. In addition, predictive maintenance in energy infrastructures is another key application of data analytics. Energy companies implement predictive maintenance to reduce downtime and extend the life of equipment. General Electric, for example, uses its Predix platform to monitor wind turbines and predict failures before they occur, reducing maintenance costs and improving system reliability.

Community energy projects also benefit from data analytics. Communities use data analytics to optimize energy consumption and manage renewable energy projects locally. The Gigha Energy Co-operative in Scotland uses data analytics to monitor the performance of its wind turbines and manage the distribution of energy among community members, optimizing usage and reducing costs. Using data visualization tools such as Power BI and community energy management platforms allows communities to understand consumption patterns and suggest actions that optimize energy use.

The analysis of large volumes of data using advanced tools and analytical methods is essential for the optimization of renewable energy projects. From generation optimization and demand management to predictive maintenance and grid integration, data analytics provides valuable insights that improve operational efficiency, reduce costs, and increase the reliability and sustainability of energy systems. The implementation examples demonstrate the enormous potential of data analytics to transform the management and operation of renewable energy projects around the world.

Chapter 7: The Nuclear Energy Debate

Introduction to Nuclear Energy

Nuclear energy is a form of energy released during nuclear reactions, either by fission, which is the splitting of heavy atomic nuclei, or fusion, which is the joining of light atomic nuclei. This energy has had a significant impact on electricity generation since its discovery. Initial discoveries in the field of nuclear energy date back to 1896 when Henri Becquerel discovered the phenomenon of radioactivity. This finding was followed by the pioneering work of Marie and Pierre Curie, who studied radioactive materials and laid the foundations for the understanding of nuclear reactions. In 1938, Otto Hahn and Fritz Strassmann, with the theoretical collaboration of Lise Meitner and Otto Frisch, discovered the nuclear fission of uranium, demonstrating that it was possible to release a large amount of energy by splitting atomic nuclei. During World War II, the Manhattan Project used nuclear fission to develop the first nuclear weapons. Although initially the use was military, this project also promoted research on peaceful applications of nuclear energy. In 1954, the Obninsk nuclear plant in the Soviet Union became the first to generate electricity from nuclear power for an electrical grid, ushering in the era of civilian nuclear power.

Nuclear fission occurs when a heavy nucleus, such as uranium-235 or plutonium-239, splits into two lighter nuclei, releasing a large amount of energy in the form of heat. This reaction also releases neutrons, which can induce fission in other nuclei, creating a controlled chain reaction in nuclear reactors. Nuclear fusion, on the other hand, involves the joining of two light nuclei, such as those of hydrogen, to form a heavier nucleus, releasing energy in the process. Fusion is the source of energy for the Sun and stars, and while it holds great promise, controlling fusion on Earth for electricity generation still faces significant technological challenges. In a nuclear fission reactor, the chain reaction is controlled by neutron-absorbing materials, known as moderators and control rods, to maintain a constant and safe fission rate.

Nuclear reactors are devices designed to control nuclear chain reactions and use the released energy to generate electricity. There are several types of nuclear reactors, each with different characteristics and operating mechanisms. The Pressurized Water Reactor (PWR) is the most common type of reactor in the world and uses light water as a moderator and coolant. High-pressure water circulates through the reactor core, where it is heated by fission and transfers its heat to a steam generator via a heat exchanger. The steam generated drives a turbine connected to an electric generator. Boiling Water Reactors (BWRs) are similar to PWRs, but the water in the reactor core boils and turns directly into steam that is directed to a turbine to generate electricity.

The Heavy Water Reactor (CANDU) uses heavy water as a moderator and coolant and can use natural uranium as fuel. Heavy water circulates through the core, moderating the neutrons and heating up, transferring the heat to a steam generator. Graphite-Gas reactors, such as AGRs and RBMKs, use graphite as a moderator and gas as a coolant. The gas circulates through the core, is heated by fission, and transfers its heat to a steam generator to produce electricity. Ball Bed Reactors (HTGRs) are high-temperature reactors that use graphite and helium-coated fuel spheres as coolant. The fuel spheres contain uranium particles and are coated with graphite, which moderates neutrons, while helium circulates through the reactor extracting the heat generated.

Molten Salt Reactors (MSRs) use a mixture of molten salts as fuel and coolant. Nuclear fuel dissolves in molten salts, which circulate through the reactor core, transferring the heat generated by fission to a heat exchanger to generate steam to drive a turbine. Although they are still in the experimental phase, there are research projects in the United States and China that explore this technology. From its initial discoveries to becoming a major source of electricity generation, nuclear power has evolved significantly. Different types of nuclear reactors offer different advantages and challenges, each adapted to different needs and technological contexts. Understanding how it works is essential to make the most of its potential and ensure its safe and efficient operation.

Nuclear Energy as Renewable Energy

Considering nuclear energy as a renewable energy source is a matter of debate, but there are several arguments and technological advances that support the idea that nuclear energy can play a crucial role in a sustainable energy future. During their operation, nuclear power plants emit negligible amounts of greenhouse gases compared to power plants that burn fossil fuels. This makes nuclear power a viable option for reducing carbon emissions and combating climate change. According to the Intergovernmental Panel on Climate Change (IPCC), nuclear power has one of the lowest carbon footprints per kilowatt-hour (kWh) generated, comparable to wind and solar power. Nuclear fuels, such as uranium and thorium, have an extremely high energy density, which means that a small amount of nuclear material can produce a large amount of energy. This high energy density allows nuclear plants to generate large amounts of electricity continuously and reliably, unlike intermittent renewable sources such as solar and wind. This is particularly beneficial for meeting base demand for electricity.

The known reserves of uranium and thorium are sufficient to supply nuclear plants for many decades, even centuries, with the use of advanced fuel recycling and reprocessing technologies. The abundance and longevity of these fuels suggest that nuclear power can be a long-term sustainable energy source, complementing other renewable sources and helping to ensure energy security. In addition, innovations in nuclear technology, such as fourth-generation reactors and small modular reactors (SMRs), are improving the safety, efficiency, and sustainability of nuclear power. Fourth-generation reactors use technologies such as gas, molten salt, and liquid metal cooling, which improves thermal efficiency and reduces the risk of serious accidents. These reactors can use alternative fuels such as thorium and recycle nuclear waste, making the process more sustainable.

Small modular reactors (SMRs) are nuclear reactors of smaller size and capacity compared to traditional reactors, designed to be mass-produced and assembled on-site. SMRs offer flexibility in the location and scaling of power generation capacity. Their modular design reduces upfront costs and construction time, making them attractive to smaller communities and developing countries. In addition, their enhanced safety features and reduced operational risk make them a safe and viable option. Another significant development in the field of nuclear energy is research on nuclear fusion energy. Nuclear fusion is the process that powers the Sun, where light nuclei combine to form a heavier core, releasing large amounts of energy. Unlike fission, fusion does not produce long-lived radioactive waste and has a lower risk of nuclear accidents. If commercialized, nuclear fusion could provide a virtually limitless source of clean and safe energy. Projects such as ITER (International Thermonuclear Experimental Reactor) in France are advancing the research and development of this technology, hoping to make it viable for large-scale power generation.

Nuclear fusion has the potential to revolutionize power generation, eliminating nuclear waste problems and providing a sustainable, carbon-free source of energy. This technological advance, along with improvements in nuclear fission, suggests that nuclear power can play a crucial role in the transition to a cleaner and more sustainable energy system. Although nuclear energy faces challenges and controversies, it makes several strong arguments in favor of its consideration as a renewable energy source. Its low greenhouse gas emissions, high energy density, fuel longevity, and current and future technological innovations position it as a viable option to contribute to a sustainable and carbon-free energy future. Nuclear energy, with its current capabilities and future potential, offers a promising avenue to address the challenges of climate change and the growing global demand for energy, complementing other renewable sources in a diversified and sustainable energy mix.

Against Nuclear Energy as Renewable Energy

Nuclear energy is a subject of intense debate, especially in the context of considering it as a renewable energy source. This discussion focuses on several arguments that question the viability and safety of nuclear energy compared to other forms of renewable energy. One of the main problems is the management and storage of radioactive waste. This waste, generated during the nuclear fission process in reactors, varies in its level of radioactivity and in the time it remains dangerous. The safe management and storage of this waste represents significant challenges, as high-radioactivity waste requires isolation for thousands of years, posing technical and ethical issues for future generations. Proposed solutions include storage in deep geological reservoirs, such as the one proposed at Yucca Mountain in the United States, and the recycling and reprocessing of nuclear fuel to reduce the volume and radioactivity of the waste. However, these solutions face technical, economic, and publicly accepted challenges.

Risks associated with nuclear waste include the possibility of radioactive leakage, environmental contamination, and the threat of misuse or terrorism. These risks require strict security measures and long-term management, which increases the costs and complexity of nuclear programs. Proposed solutions to mitigate these risks include developing new secure storage technologies, improving surveillance and security systems, and implementing strict policies and regulations to ensure the safety of waste deposits. The history of nuclear accidents has left an indelible mark on the public perception of nuclear energy. The Chernobyl accident in 1986, which resulted in an explosion and fire at the Chernobyl nuclear reactor in Ukraine, released large amounts of radiation into the environment. This caused the evacuation of thousands of people and contaminated vast areas, resulting in a high incidence of cancers and other radiation-related diseases, as well as long-term economic and environmental impacts. Another significant disaster was the Fukushima accident in 2011, where an earthquake followed by a tsunami caused the cooling system of the reactors at the Fukushima Daiichi nuclear plant in Japan to fail. This incident caused the meltdown of the core in several reactors and the release of radioactivity into the environment, leading to the evacuation of thousands of people and generating a global reassessment of nuclear safety and energy policy.

Despite these events, modern nuclear plants implement numerous safety measures designed to prevent accidents and mitigate their effects should they occur. These measures include redundant cooling systems, concrete containments to prevent the release of radiation, and well-defined and regularly tested emergency protocols. Emergency protocols include evacuation plans, iodine distribution to protect the thyroid from radiation, and communication systems to inform the public and coordinate responses with local and national authorities. The initial costs of building and maintaining nuclear power plants are considerably high due to design and safety requirements, technological complexity, and long development times. In addition, maintenance, fuel, and waste management add ongoing costs. Nuclear plants also require safe decommissioning at the end of their useful life, which entails significant additional costs. Compared to other energy sources, the initial costs of nuclear plants are generally higher, although operating costs can be competitive once the plant is operational.

Compared to the costs of other renewable energy sources, such as solar and wind power, nuclear facilities face economic challenges. Solar and wind installations have lower upfront costs and shorter construction times than nuclear plants. The operating costs of solar and wind power are low, as they do not require fuel and maintenance is relatively straightforward. Technological improvements and economies of scale have significantly reduced the costs of solar and wind energy, making them increasingly competitive. Hydropower also has high upfront costs like those of nuclear power, especially for large dams and pumped storage projects. However, the operating costs of hydropower are generally low, and hydropower plants have a long lifespan. Hydropower is a reliable and well-established renewable source, but its expansion is limited by the availability of suitable sites.

Although nuclear power has significant advantages, it also faces critical challenges related to waste management, safety, and economic costs. Radioactive waste and the risks of nuclear accidents present complex problems that require long-term solutions and rigorous management. In addition, high upfront and maintenance costs, coupled with economic considerations, can make nuclear power less attractive compared to other cheaper, rapidly deployable renewable energy sources. The need to address these issues is essential to determine the future of nuclear energy in the context of the transition to a more sustainable and secure energy system.

Global Perspectives on Nuclear Energy

Nuclear energy plays a varied role in the energy policies of different countries, reflecting a mix of expansion and phase-out strategies according to national priorities. France, for example, is one of the biggest proponents of nuclear power in the world, generating about 70% of its electricity through nuclear plants. The country has invested significantly in nuclear infrastructure and advanced technologies, such as third- and fourth-generation reactors, and is exploring small modular reactors (SMRs) to diversify its nuclear capacity. This dependence has allowed France to maintain low carbon emissions in its energy sector, contributing substantially to its climate goals.

China, on the other hand, is rapidly expanding its nuclear capacity as part of its strategy to reduce reliance on coal and decrease greenhouse gas emissions. The Chinese government has set ambitious plans to increase nuclear capacity, with multiple reactors under construction and the development of advanced technologies such as fourth-generation reactors and SMRs. This expansion seeks to diversify the country's energy mix, improve energy security and contribute to the reduction of carbon emissions, which is vital for its sustainable growth.

Russia is also a key player in nuclear energy, with an active reactor construction policy both domestically and internationally. Through Rosatom, its state nuclear power corporation, Russia is not only increasing its domestic capacity but also exporting technology and building reactors in other countries. This approach not only strengthens its domestic electricity generation but also expands its geopolitical influence by exporting nuclear technology, making nuclear energy a fundamental pillar of its energy and foreign relations strategy.

On the contrary, countries such as Germany and Switzerland are phasing out nuclear power due to safety and sustainability concerns. Following the Fukushima disaster in 2011, Germany decided to shut down all its nuclear plants by 2022 as part of its Energiewende policy, which focuses on increasing capacity from renewables such as solar and wind and improving energy efficiency. Although this phase-out has boosted the development of renewables, it also poses challenges in terms of maintaining grid stability and meeting carbon emission reduction targets.

Switzerland also decided to phase out nuclear power after Fukushima, although the process is slower compared to Germany. The country plans to shut down its nuclear plants at the end of their useful life without building new ones, focusing on increasing renewable capacity and improving energy efficiency. This policy reflects public safety concerns, and Switzerland is working to balance its energy mix with more renewables and efficiency improvements.

The social acceptance of nuclear energy varies significantly between different countries and regions, influenced by historical, cultural and political factors. Various polls and studies show a range of opinions on nuclear power, from enthusiastic support to vehement opposition. In countries like France and Russia, where nuclear power is an integral part of energy infrastructure, acceptance tends to be higher. In contrast, in countries such as Germany and Japan, nuclear disasters have generated strong public opposition.

The perception of safety and the associated risks play a crucial role in society's acceptance or rejection of nuclear energy. Nuclear incidents such as Chernobyl and Fukushima have left a lasting impact on public perception, fueling fears of accidents and radioactive waste management. Trust in regulatory institutions and operating companies is also a determining factor; Countries with strong and transparent institutions tend to have greater public acceptance. In addition, the perception of nuclear energy as a solution to the climate crisis, capable of providing steady, low-carbon energy, may increase its acceptance. Economic benefits, such as job creation and energy stability, are also positive factors.

The level of knowledge and education about nuclear energy significantly influences public attitudes. A better understanding of the benefits and risks can lead to a more balanced opinion. The media and activist movements play a crucial role in shaping public opinion. Negative media coverage and anti-nuclear campaigns can reduce social acceptance, while balanced, fact-based reporting can improve it.

The global perspective on nuclear energy is diverse and complex. While some countries continue to invest in nuclear energy as a key solution to their energy needs and climate goals, others are phasing it out due to safety concerns and public acceptance. Public opinion on nuclear energy is influenced by a variety of factors, including perception of safety, trust in institutions, economic and environmental benefits, and education and knowledge about nuclear technology. The various approaches reflect the different priorities and challenges faced by each country in its quest for a sustainable energy future.

The Future of Nuclear Energy

The future of nuclear energy is shaping up with a series of innovations and developments that seek to make this energy source safer, more efficient and sustainable. Among the most promising advances are fourth-generation reactors. These reactors represent a new class of nuclear designs that promise to be safer, more efficient, and more sustainable. They include types such as the sodium-cooled fast reactor, the molten salt reactor, and the high-temperature gas reactor. These designs are optimized to make better use of fuel, reduce nuclear waste, and increase operational safety through intrinsic characteristics that prevent accidents. The implementation of fourth-generation reactors could solve many of the problems associated with current nuclear power, making this energy source more palatable to the public and more viable in the long term.

In addition to fourth-generation reactors, small modular reactors (SMRs) are gaining attention. SMRs are smaller-scale nuclear reactors, designed to be safer and more economical. They can be mass-produced and assembled on-site, reducing construction costs and implementation times. These reactors offer flexibility in power generation and can be used in remote locations or in combination with other renewable energy sources. Its advanced safety features reduce the risk of accidents and improve public acceptance. The adoption of SMRs could democratize nuclear energy, allowing more countries and regions to access this technology with lower risks and costs.

Another crucial area of research is nuclear fusion energy. Nuclear fusion is the process that powers the Sun and stars, where light cores combine to form a heavier core, releasing an enormous amount of energy. Current research seeks to replicate this process on Earth to generate electricity safely and sustainably. International projects such as ITER (International Thermonuclear Experimental Reactor) in France are advancing in the research and development of nuclear fusion. These projects are designed to demonstrate the technical and economic feasibility of fusion as an energy source. If commercialized, nuclear fusion could revolutionize power generation, providing a nearly limitless, carbon-free source without the long-lived radioactive waste problems associated with nuclear fission.

The integration of nuclear energy with other renewable energy sources is also a key aspect of the energy future. Nuclear power can complement intermittent renewable energy sources such as solar and wind, providing a constant and reliable source of electricity that can balance the variability of renewables. Hybrid models of nuclear and renewable plants can leverage the strengths of each technology, improving grid stability and reducing carbon emissions. For example, a nuclear plant can operate at baseload while solar and wind power supply additional demand during production peaks. The combination of nuclear and renewables can result in a more robust and resilient energy system, capable of providing continuous electricity and reducing dependence on fossil fuels.

A diversified approach to the energy mix is critical for a sustainable future. A sustainable energy mix combines multiple energy sources, including nuclear, solar, wind, hydroelectric and biomass, to maximize reliability and minimize carbon emissions. A diversified approach reduces vulnerability to fluctuations from a single energy source and takes advantage of each technology. Nuclear power provides a firm base of generation, while renewables cover variable demand and reduce the carbon footprint. Countries such as France and China are exploring energy mix models that integrate nuclear energy with renewables, taking advantage of the stability of nuclear and the sustainability of renewables.

The development of smart infrastructure is essential for the successful integration of nuclear energy and renewables. Smart infrastructure includes smart grids, energy storage, and demand management systems. These technologies can optimize energy use, reduce losses, and improve responsiveness to fluctuations in generation and demand, making the energy mix more efficient and sustainable. Integrating these advanced technologies can transform the way energy is managed and distributed, improving reliability and reducing operating costs.

In short, the future of nuclear energy is marked by innovations and developments that promise to make it safer, more efficient and compatible with other renewable energy sources. Research into advanced reactors and nuclear fusion power could transform power generation, providing sustainable, low-carbon solutions. The integration of nuclear energy with renewable technologies and the development of a diversified energy mix are key strategies for a sustainable, resilient and carbon-free energy future. Not only do these advances have the potential to solve the current problems associated with nuclear power, but they could also set new standards for clean and safe energy generation in the 21st century.

Conclusions on Nuclear Energy

The chapter on nuclear energy has provided a detailed and comprehensive overview of this energy source, highlighting both its benefits and challenges. Nuclear energy, based on the fission of heavy atoms such as uranium and plutonium, has been a key element in the generation of electricity since the mid-twentieth century. This technology harnesses the energy released during nuclear fission to produce large amounts of electricity efficiently. Throughout history, several types of nuclear reactors have been developed, each with specific characteristics and applications, such as pressurized water reactors (PWRs), boiling water reactors (BWRs), heavy water reactors (CANDU), graphite-gas reactors (AGR and RBMK), and ball bed reactors (HTGRs).

A highlight of the chapter is the consideration of nuclear energy as a renewable energy source due to its low greenhouse gas emissions during operation and its high energy density. This capacity allows for the generation of large amounts of electricity, which is crucial in the fight against climate change. Recent innovations, such as fourth-generation reactors and small modular reactors (SMRs), are designed to improve the safety and efficiency of nuclear plants, addressing some of the most persistent concerns around this technology.

However, nuclear energy is not without its challenges and controversies. Radioactive waste management remains a significant problem, with waste requiring safe storage for thousands of years. The risks of nuclear accidents, such as those at Chernobyl and Fukushima, have left an indelible mark on public opinion and have led to debates about the viability and safety of nuclear energy. In addition, the high upfront costs of building and maintaining nuclear plants are economic obstacles that must be considered.

Policies and strategies towards nuclear energy vary considerably between countries. While France, China, and Russia are investing in and expanding their nuclear capabilities, others such as Germany and Switzerland are phasing out this energy source due to security concerns and social acceptance. This contrast reflects the diversity of approaches and priorities in global energy policy.

The future of nuclear energy looks promising thanks to technological innovations. Fourth-generation reactors and small modular reactors offer solutions that can make nuclear power safer and more efficient. Research in nuclear fusion energy, which seeks to replicate the process that powers the Sun, promises an almost limitless source of clean and safe energy, although it still faces significant technical challenges.

The integration of nuclear energy with other renewable energy sources is a key strategy for a sustainable energy mix. Nuclear power can provide a constant and reliable source of electricity that complements the intermittent nature of renewables such as solar and wind. This hybrid approach can improve grid stability and reduce carbon emissions, leveraging the strengths of each technology.

In the realm of energy policy, it is crucial that governments implement strict regulations and safety protocols to minimize the risks associated with nuclear power. Financial support for research and development of advanced nuclear technologies, together with effective strategies for waste management, is critical to advancing this field. Transparency and trust in regulatory institutions also play an essential role in the public acceptance of nuclear energy.

In terms of future research, it should focus on the development of safer and more efficient nuclear technologies, as well as on the integration of nuclear energy with other renewable sources. Education and effective communication are essential to address public concerns and foster an informed understanding of the benefits and risks of nuclear energy.

In summary, nuclear power has the potential to play a significant role in the transition to a more sustainable and carbon-free global energy system. Technological innovations and supportive policies can facilitate their integration with other renewable energy sources, contributing to a cleaner, safer and more sustainable energy future. Continuous research and the development of effective communication strategies are essential to ensure that nuclear energy can effectively contribute to the global energy transition.

Conclusion

In short

Throughout this book, we have examined a wide range of topics related to renewable energy and the technological innovations that are transforming this sector. Renewable energy has come a long way since its inception and currently plays a crucial role in the transition to a more sustainable energy system. Technologies such as solar, wind, hydroelectric and biomass are increasingly being integrated into the electricity grids of many countries, thanks to significant advances in efficiency and cost reduction.

One of the highlights is the evolution of photovoltaic technologies. Solar panels have improved significantly in terms of efficiency and cost, with new materials such as perovskites promising to further increase the accessibility of solar energy. These technologies are being used in a variety of applications, from small residential systems to large industrial-scale solar plants, demonstrating their versatility and effectiveness.

Wind energy has also seen notable advances, especially in the design and efficiency of onshore and offshore turbines. Floating wind turbines are expanding the reach of wind energy into new areas, providing a steady and potent source of renewable electricity. These developments have demonstrated the viability and efficiency of wind power around the world, from onshore projects in Europe to offshore wind farms in Asia.

Energy storage is another critical component for the integration of renewable energy. Advanced batteries, including lithium-ion, solid-state and redox flow batteries, are improving storage capacity, allowing for greater integration of renewables into the electricity grid. In addition, thermal and mechanical storage technologies, such as compressed air energy storage and flywheels, offer additional solutions for storing energy efficiently and economically.

Grid management is critical to maximizing the positive impact of renewables. Demand management systems and smart solutions use artificial intelligence and data analytics to optimize energy use and improve grid stability. The integration of renewables into the power grid includes the use of advanced technologies and energy management models to balance supply and demand, ensuring efficient and reliable operation of the energy system.

Government policies and economic incentives play a vital role in driving the development and adoption of renewable energy. Countries such as Germany, China, and Denmark have implemented successful policies that have significantly promoted renewable energy. Grants, tax credits, and other incentives have been crucial in reducing upfront costs and encouraging investment in renewable technologies, demonstrating the importance of government support in this sector.

Innovative financing strategies are also facilitating investment in renewable energy projects. Business models such as power purchase agreements (PPAs), crowdfunding, and green bonds are helping to finance these projects. Successful examples of projects financed with these models demonstrate their viability and potential to drive the energy transition.

The optimization and maintenance of energy systems have improved thanks to the use of artificial intelligence and the analysis of large volumes of data. These technologies are transforming the management of energy systems, improving operational efficiency and reliability. Predictive maintenance, which makes it possible to predict failures and carry out proactive interventions, is reducing costs and downtime, ensuring a more efficient operation of energy infrastructures.

Nuclear power, while facing significant challenges, remains an important source of low-carbon electricity generation. Advances in safer and more efficient reactors, and research into nuclear fusion energy, promise a more sustainable future for this technology. The integration of nuclear energy with other renewable sources could provide a robust and resilient energy mix, taking advantage of each technology to ensure a continuous and reliable supply of electricity.

In summary, this book has provided a comprehensive overview of the developments, challenges, and opportunities in the field of renewable energy and nuclear energy. The combination of effective policies, technological innovations, and financing strategies can drive the transition to a more sustainable and carbon-free global energy system. Collaboration between governments, businesses and civil society will be crucial to achieving these goals and ensuring a cleaner and more equitable energy future for all. Continuous research and development of new technologies will be essential to address today's challenges and make the most of the opportunities offered by renewables and nuclear energy.

What about the future?

The future of renewable energy is promising and essential to meet the challenges of climate change and ensure energy sustainability. Continuous technological advances are positioning renewables at the heart of the transition to a cleaner and more resilient global energy system. With the steady declining costs of technologies such as solar, wind, energy storage, and smart grids, the adoption of renewables will accelerate globally. This progress is supported by innovations in nuclear power, such as fourth-generation reactors and small modular reactors, as well as the emerging potential of fusion energy. These innovations promise safer and more efficient power generation, complementing renewable energy sources to create a diversified and robust energy mix that meets global energy demand in a sustainable way.

In the long term, the combination of supportive government policies, economic incentives and international collaboration will be essential to overcome the remaining technical and economic challenges. The energy transition will require strategic planning and a focus on equity, ensuring that the benefits of renewables and nuclear power are accessible to all communities, including the most vulnerable and disadvantaged. To achieve this goal, researchers, policymakers, and businesses have critical roles to play.

For researchers, technological innovation is crucial. It is imperative to continue investing in the research and development of new renewable and nuclear technologies, focusing on improving efficiency, reducing costs and increasing safety. Interdisciplinary collaboration is equally vital, bringing together experts in energy science, engineering, environmental science, and social sciences to develop comprehensive solutions to the complex challenges of the energy transition. In addition, fostering knowledge transfer between academic institutions, industry, and policymakers will accelerate the adoption of innovative technologies, contributing to a more sustainable energy system.

Policymakers should develop and maintain a clear and supportive regulatory framework that incentivizes investment in renewables and advanced nuclear technologies. This includes implementing emissions standards, feed-in tariffs, and tax credits. It is also critical to establish and strengthen rigorous regulations for nuclear safety and radioactive waste management, thereby gaining the public's trust and ensuring long-term sustainability. Energy policies must consider equity and social justice, ensuring that the benefits of the energy transition reach all communities. Support programs for job training and infrastructure development in disadvantaged areas are essential to achieve an inclusive transition.

Companies, for their part, must adopt sustainable practices and take responsibility for reducing their carbon footprint. Investing in renewables and clean technologies not only improves your public image, but also offers long-term competitive advantages. Collaboration and partnerships with governments, academic institutions, and non-governmental organizations can drive innovation and implementation of sustainable energy projects. In addition, transparency in operations and effective communication about the benefits and challenges of renewables and nuclear energy are key to increasing public acceptance and fostering stakeholder support.

In summary, the outlook for renewables and nuclear energy is encouraging, with great potential to contribute to a more sustainable and resilient global energy system. Researchers, policymakers, and businesses have crucial roles to play in this transition. Through continuous innovation, strategic policymaking, and effective collaboration, we can overcome today's challenges and create an energy future that benefits all communities and preserves the planet for future generations. Recent research has shown that with the right commitment and effective policy implementation, it is possible to achieve a successful energy transition that ensures a clean, secure, and accessible energy supply for all.

What can you do?

The transition to a sustainable energy system is not only an urgent need but a historic opportunity to build a cleaner, healthier and fairer future. Each of us, from individuals to governments to businesses, has an essential role to play in this process. The challenge of climate change and the demand for cleaner, more accessible energy call us all to action.

To readers, I invite you to educate yourself and raise awareness about renewable energies and their benefits. Educating yourself and sharing knowledge with friends, family, and communities can establish a strong foundation of support for clean energy policies and projects. In addition, it is crucial to adopt responsible energy consumption habits. Opting for sustainable products and services, reducing energy waste in our homes, and considering installing renewable energy systems, such as solar panels, are all significant steps we can take. Actively engaging in local and national initiatives that promote renewable energy is also vital. Participating in community meetings, supporting environmental organizations, and voting for policies and representatives that prioritize sustainability and energy innovation can make a noticeable difference.

To governments, I urge you to develop and implement policies that encourage research, development, and deployment of renewable technologies. It is essential to establish economic incentives, energy efficiency standards and clear targets for the reduction of carbon emissions. In addition, increasing funding for research and development of new energy technologies is crucial. Supporting academic institutions and startups that are working on innovative solutions to energy challenges can accelerate progress. International collaboration is also essential. Promoting global cooperation on renewable energy can accelerate technological development, share best practices, and increase capacity to respond to climate challenges.

To companies, I encourage them to integrate sustainability into the core of their operations. Implementing energy efficiency practices, investing in renewables, and reducing carbon emissions across the value chain are crucial steps. Fostering innovation within their organizations and collaborating with academic institutions and research centers to develop new energy technologies is vital. Participating in pilot projects and test programs that can demonstrate the viability and benefits of renewables is also critical. In addition, it is important to communicate sustainability efforts clearly and transparently. By publicly committing to renewable energy and emissions reduction targets, companies can positively influence public opinion and motivate other companies to follow suit.

The transition to a sustainable energy future is a collective effort that requires the active participation of individuals, governments and businesses. Innovation and the adoption of renewable energy are not only essential to mitigate the effects of climate change, but also represent an opportunity to create a fairer, healthier and more prosperous world. As Mahatma Gandhi said, "Be the change you want to see in the world." Let's respond to this call to action together, promoting and supporting renewable energy in all aspects of our lives and communities. With determination and collaboration, we can achieve a clean and sustainable energy future for present and future generations.

Appendices

Frequently asked questions

1. What are renewable energies?

Renewable energies are those energy sources that are obtained from inexhaustible natural resources or that are continuously regenerated, such as solar, wind, hydroelectric and biomass energy. These energy sources are considered sustainable because they are not depleted and have a lower environmental impact compared to fossil fuels.

2. What is the difference between fission and nuclear fusion?

Nuclear fission is the process in which a heavy atomic nucleus splits into two lighter nuclei, releasing a large amount of energy. This process is used in today's nuclear reactors. Nuclear fusion, on the other hand, is the process where two light nuclei combine to form a heavier nucleus, releasing energy. Fusion is the process that powers the Sun and stars and has the potential to be a virtually limitless and clean source of energy if it can be controlled on Earth.

3. What are the main advantages of renewable energies?

Renewables have several advantages, including:

- Reduction of greenhouse gas emissions.
- Diversification of the energy matrix and increase in energy security.
- Green job creation and economic development.
- Reduction of dependence on fossil fuels.
- Lower environmental impact and conservation of natural resources.

4. What challenges does nuclear energy face?

Nuclear power faces several challenges, including:

- Safe management and storage of radioactive waste.
- Risk of nuclear accidents and their impact on health and the environment.
- High upfront costs of building and maintaining nuclear plants.
- Negative public perception and lack of social acceptance in some countries.

5. How can renewables and nuclear power work together?

Renewables and nuclear power can complement each other to create a robust and sustainable energy mix. Nuclear power can provide a steady and reliable source of electricity, while renewables such as solar and wind can meet varying demand. This combination can improve the stability of the power grid and reduce carbon emissions.

6. What are small modular reactors (SMRs) and why are they important?

Small modular reactors (SMRs) are nuclear reactors of smaller size and capacity than traditional reactors. They are important because they offer advantages such as:

- Increased safety due to intrinsically safe design features.
- Lower upfront costs and shorter construction times.
- Flexibility in location and the ability to be assembled in series.
- Potential to be used in remote areas or in combination with other renewable energy sources.

7. What innovations are solar and wind technologies improving?

Some recent innovations in solar and wind technologies include:

- High-efficiency solar panels with new materials such as perovskites.

- Wind turbines with advanced designs that increase efficiency and generation capacity.
- Development of energy storage technologies that allow better management of the intermittency of renewable energies.
- Hybrid systems that combine solar, wind and storage to optimize energy generation and use.

8. What is energy demand management and why is it important?

Energy demand management involves the use of strategies and technologies to influence end-users' energy consumption. It is important because it helps balance supply and demand in the electricity grid, improves energy efficiency, reduces costs, and facilitates the integration of renewable energy into the grid.

9. How can renewable energy projects be financed?

Renewable energy projects can be financed through several mechanisms, including:

- Grants and tax credits offered by governments.
- Green bonds that attract sustainable investments.
- Power purchase agreements (PPAs) that ensure a long-term revenue stream.

- Crowdfunding and community participation to finance local projects.
- Investment funds specialising in clean energy.

10. What role do government policies play in the adoption of renewable energy?

Government policies are crucial for the adoption of renewable energy. They can provide economic incentives, set clean energy targets, implement energy efficiency standards, and regulate the energy market to encourage investment in renewable technologies. A favourable regulatory framework and constant support are essential to drive the transition to a sustainable energy system.

11. What is energy storage and why is it crucial for renewable energy?

Energy storage refers to technologies and systems that store energy for later use. It is crucial for renewables because it allows excess energy generated during periods of high production (e.g. sunny or windy days) to be stored and that energy released when demand is high, or production is low. This helps to stabilize the power grid and ensures a constant and reliable power supply.

12. What are smart grids and how do they help in the integration of renewable energies?

Smart grids are advanced electrical systems that use digital technology to monitor and manage the flow of electricity efficiently. They assist in the integration of renewable energy by real-time management of supply and demand, improving operational efficiency, reducing losses, and facilitating energy storage and demand response.

13. What are the main obstacles to the mass adoption of renewable energy?

The main obstacles include:

- Intermittency and variability in renewable energy production.
- Lack of adequate energy storage and transmission infrastructure.
- High initial installation and development costs.
- Inconsistent regulatory barriers and policies.
- Social resistance and lack of acceptance in some communities.

14. What is the circular economy and how is it related to renewable energy?

The circular economy is an economic model that seeks to minimize waste and make more efficient use of resources through the recycling, reuse, and regeneration of materials. It is related to renewable energy in that it promotes sustainable and efficient practices that can reduce dependence on non-renewable resources and minimize the environmental impact of power generation.

15. How can local communities benefit from renewable energy projects?

Local communities can benefit from renewable energy projects through:

- Creation of local jobs in the construction, operation and maintenance of facilities.
- Reduced energy costs and increased energy independence.
- Additional income through participation in community projects and the sale of surplus energy.
- Improved air quality and reduced environmental pollution.
- Strengthening the economic and energy resilience of the community.

16. What is a power purchase agreement (PPA) and how does it work?

A power purchase agreement (PPA) is a contract between a power producer and a buyer (usually a company or utility) in which the buyer agrees to purchase the electricity generated at a fixed price for a specified period. PPAs provide revenue predictability for renewable energy project developers and reduce the financial risks associated with price variability in the electricity market.

17. What role do technological innovations play in reducing the costs of renewable energies?

Technological innovations play a crucial role in reducing the costs of renewable energy by:

- Improving energy conversion efficiency in technologies such as solar panels and wind turbines.
- Development of cheaper and more durable materials.
- Optimization of manufacturing and construction processes.
- Advances in energy storage technologies and smart grids.
- Facilitate the integration and management of renewable energies in the electricity grid.

18. What is geothermal energy and how is it used?

Geothermal energy is the energy obtained from the Earth's internal heat. It is mainly used for electricity generation and for heating and cooling applications. In geothermal plants, heat is extracted from the subsurface through wells and used to generate steam that drives turbines connected to electric generators. It is also used in geothermal heating systems for residential buildings and districts.

19. How can artificial intelligence (AI) improve the efficiency and management of renewable energy systems?

Artificial intelligence can improve the efficiency and management of renewable energy systems by:

- Accurate prediction of power generation based on weather data and historical patterns.
- Optimization of the operation and maintenance of facilities through data analysis and anomaly detection.
- Real-time energy demand and supply management to balance the electricity grid.
- Process automation and data-driven decision-making to improve operational efficiency.

20. What are green bonds and how do they support the development of renewable energy projects?

Green bonds are debt instruments issued to finance projects that have environmental benefits, such as renewable energy projects, energy efficiency, clean transportation, and waste management. The proceeds generated by green bonds go exclusively to these projects, providing investors with a way to support sustainable initiatives while earning a financial return. Green bonds help mobilize capital for the development and expansion of clean and sustainable energy projects.

Glossary of Technical Terms

- **Renewable Energy**: Energy obtained from natural sources that are constantly regenerated, such as solar, wind, hydroelectric and biomass.
- **Nuclear Energy**: Energy released during the fission or fusion of atomic nuclei, used mainly in the generation of electricity.
- **Nuclear Fission**: The process in which a heavy atomic nucleus splits into two or more light nuclei, releasing a large amount of energy.
- **Nuclear Fusion**: The process in which two light nuclei combine to form a heavier nucleus, releasing energy. It is the process that powers the Sun and the stars.
- **Nuclear Reactor**: A device in which nuclear chain reactions are controlled to produce energy in the form of heat.
- **Pressurized Water Reactor (PWR):** A type of nuclear reactor that uses high-pressure water as a moderator and coolant.
- **Boiling Water Reactor (BWR):** A type of nuclear reactor in which water boils in the core of the reactor and the steam produced is used to generate electricity.

- **Heavy Water Reactor (CANDU):** A type of nuclear reactor that uses heavy water as a moderator and coolant, and can use natural uranium as fuel.
- **Small Modular Reactors (SMRs):** Nuclear reactors of smaller size and capacity that can be mass-produced and assembled on site, designed to be safer and more economical.
- **Photovoltaic Solar Energy:** Technology that converts sunlight directly into electricity through the use of solar panels composed of photovoltaic cells.
- **Concentrated Solar Energy (CSP):** Technology that uses mirrors or lenses to concentrate a large area of sunlight onto a small receiver, generating heat that is converted into electricity.
- **Wind turbine**: A device that converts the kinetic energy of the wind into electrical energy by using blades that rotate a generator.
- **Offshore Wind Energy:** Generation of electricity by wind turbines located at sea, where the winds are generally stronger and more constant.
- **Energy Storage:** Technologies and systems that store energy for later use. Examples include batteries, compressed air energy storage (CAES), and flywheels.

- **Lithium-Ion Batteries:** A type of rechargeable battery commonly used in electronic devices and electric vehicles, known for its high energy density and long lifespan.
- **Smart Grids:** Electricity grids that use advanced digital technology to monitor and manage the flow of electricity efficiently, improving the reliability and sustainability of the system.
- **Artificial Intelligence (AI):** A field of computer science that focuses on creating systems capable of performing tasks that require human intelligence, such as learning, decision-making, and problem-solving.
- **Predictive Maintenance:** Maintenance strategy that uses advanced data and analytics to predict when equipment failures are likely to occur, enabling preventative interventions.
- **Big Data Analytics:** The process of examining large volumes of data to discover patterns, trends, and associations that may be useful for decision-making.
- **Tax Credits:** Financial incentives that allow taxpayers to reduce the amount of their taxes, used to promote investments in renewable energy and clean technologies.

- **Grants**: Direct financial contributions from the government to support specific projects, such as the installation of renewable energy technologies.
- **Green Bonds:** Debt instruments issued to finance projects that have environmental benefits, such as renewable energy and sustainability projects.
- **Compressed Air Energy Storage (CAES):** Energy storage technology that uses compressed air in underground cavities or high-pressure tanks, releasing the air to generate electricity when needed.
- **Flywheels**: Devices that store kinetic energy in a rotating rotor, releasing energy when needed by decelerating the rotor.
- **Energy Policy**: A set of decisions and actions taken by a government to manage and regulate the production, distribution, and consumption of energy in a country or region.
- **Circular Economy**: Economic model that seeks to minimize waste and make more efficient use of resources, through the recycling, reuse and regeneration of materials.
- **Energy Mix**: A combination of different energy sources used to meet the electricity demand of a region or country, including renewable, nuclear, and fossil fuels.

- **Radioactive Waste**: Materials that contain radioactive isotopes and are produced as byproducts of nuclear reactions. They require safe management and storage due to their potential health and environmental hazards.
- **Energy Demand Management**: Strategies and technologies used to influence end-users' energy consumption, with the aim of balancing supply and demand in the electricity grid.
- **Energy Resilience**: The ability of an energy system to adapt and recover from disruptions and challenges, ensuring a continuous and reliable supply of electricity.

Recommended Reading List

1. "Sustainable Energy – Without the Hot Air" by David JC MacKay

- Description: This book provides a detailed and accessible analysis of the different sustainable energy sources and their actual capabilities, with a focus on clear data and calculations.

- Why read it: It offers a solid, data-driven understanding of the possibilities and limitations of renewable energy.

2. "The Switch: How solar, storage and new tech means cheap power for all" by Chris Goodall

- Description: Explore how solar and energy storage technologies are revolutionizing the energy market, making clean energy more accessible and affordable.

- Why read it: It provides an optimistic view of the future of solar energy and energy storage innovations.

3. **"Power to the People: How the Coming Energy Revolution Will Transform an Industry, Change Our Lives, and Maybe Even Save the Planet" by Vijay V. Vaitheeswaran**

- Description: Analyzes how new energy technologies and policies are transforming the energy sector and addresses the challenges and opportunities of this transition.

- Why read it: It offers a global perspective on the ongoing energy revolution and its potential impact.

4. **"Renewable Energy: Power for a Sustainable Future" by Stephen Peake and Joe Smith**

- Description: This book is a comprehensive resource covering all aspects of renewable energy, from basic principles to practical applications and policies.

- Why read it: It's a comprehensive and educational guide that covers a wide range of renewable energy technologies and issues.

5. "The Future of Fusion Energy" by Jason Parisi and Justin Ball

- Description: An accessible and well-informed introduction to nuclear fusion, its principles, technological advances, and the path to its commercial viability.

- Why read it: It provides a clear view of developments in fusion energy and its potential for the energy future.

6. "Energy Transitions: Global and National Perspectives" by Vaclav Smil

- Description: Examines the history of energy transitions and offers perspectives on the future of global and national energy.

- Why read it: It offers historical context and deep insight into how and why energy transitions occur.

7. "Nuclear Energy: Principles, Practices, and Prospects" by David Bodansky

- Description: Provides a comprehensive understanding of nuclear energy, including scientific principles, current technologies, and future prospects.

- Why read it: It is a detailed and well-documented reference source on all aspects of nuclear energy.

8. "Reinventing Fire: Bold Business Solutions for the New Energy Era" by Amory B. Lovins

- Description: It presents a detailed strategy for a transition to a sustainable energy future, without the need for coal, oil and nuclear power.

- Why read it: It offers practical and bold solutions to transform the global energy system towards sustainability.

9. "Smart Grid: Integrating Renewable, Distributed & Efficient Energy" by Fereidoon P. Sioshansi

- Description: Examines the development and implementation of smart grids that integrate renewable energy, distributed technologies, and energy efficiency.

- Why read it: It provides an in-depth analysis of how smart grids can facilitate the transition to a more sustainable and efficient energy system.

10. "Clean Disruption of Energy and Transportation" by Tony Seba

- Description: Argues that the convergence of disruptive technologies such as electric vehicles, solar energy, and energy storage will completely transform energy and transportation.

- Why read it: It offers a bold and compelling vision for how emerging technologies are shaping the future of energy and transportation.

Helpful Resources and Tools

Online Resources

1. International Renewable Energy Agency (IRENA)

- Website: www.irena.org

- Description: Provides data, reports, and analysis on renewable energy, including global statistics and market trends.

2. International Energy Agency (IEA)

- Website: www.iea.org

- Description: Provides detailed reports, data, and projections on the global energy sector, with a focus on the energy transition and energy policies.

3. Renewable Energy World

- Website: www.renewableenergyworld.com

- Description: Provides news, articles, and analysis on renewable energy technologies, projects, and policies.

4. National Renewable Energy Laboratory (NREL)

- Website: www.nrel.gov

- Description: Provides research, data, and tools on renewable energy technologies and energy efficiency.

5. Global Wind Energy Council (GWEC)

- Website: www.gwec.net

- Description: Provides reports and statistics on the wind energy industry globally, including market trends and policies.

6. Solar Energy Industries Association (SEIA)

- Website: www.seia.org

- Description: Provides resources, reports, and news about the solar industry in the United States, including policies and market trends.

Tools & Software

1. HOMER Energy

- Website: www.homerenergy.com

- Description: Software for the modeling and optimization of hybrid energy systems that combines renewable energies and storage.

2. SAM (System Advisor Model)

- Website: sam.nrel.gov

- Description: Tool developed by NREL for modelling and financial analysis of renewable energy projects, including solar, wind and biomass.

3. RETScreen

- Website: www.nrcan.gc.ca/maps-tools-publications/tools/retscreen/7465

- Description: Software for the analysis of clean energy projects, including feasibility assessment, performance analysis, and financial analysis.

4. PVsyst

- Website: www.pvsyst.com

- Description: Software for the sizing and simulation of photovoltaic systems, used for the analysis and design of solar installations.

5. WindPRO

- Website: www.emd.dk/windpro

- Description: Software tool for the planning and analysis of wind energy projects, including wind farm design and wind resource assessment.

6. EnergyPLAN

- Website: www.energyplan.eu

- Description: Simulation tool for the modelling of sustainable energy systems, used to analyse the integration of different energy sources and technologies.

Publications and Databases

1. World Energy Outlook (IEA)

- Description: Annual report that provides detailed analysis and projections on the situation and trends of the global energy sector.

- Website: www.iea.org/weo

2. Renewables Global Status Report (REN21)

- Description: Annual report that provides a global view of the renewable energy market, industry, investment and policies.

- Website: www.ren21.net/reports/global-status-report

3. BP Statistical Review of World Energy

- Description: Annual publication that provides data and analysis on energy consumption, production, and trends around the world.

- Website: www.bp.com/statisticalreview

4. Database of State Incentives for Renewables & Efficiency (DSIRE)

- Description: Comprehensive database of U.S. state and federal incentives and policies for renewables and energy efficiency.

- Website: www.dsireusa.org

5. Global Energy Statistical Yearbook

- Description: A database that provides statistics and analysis on global energy consumption, production, and trade.

- Website: yearbook.enerdata.net

Professional Organizations and Networks

1. International Solar Energy Society (ISES)

- Website: www.ises.org

- Description: Global network of solar energy professionals and experts, promoting the research, development and application of solar technologies.

2. International Association for Energy Economics (IAEE)

- Website: www.iaee.org

- Description: Professional organization dedicated to advancing knowledge about energy economics, through research, education, and information exchange.

3. American Wind Energy Association (AWEA)

- Website: www.awea.org

- Description: National Wind Industry Association in the United States, which promotes wind energy and supports its members through advocacy, research, and education.

4. European Renewable Energy Council (EREC)

- Website: www.erec.org

- Description: Organization that represents renewable energy industries in Europe and promotes sustainable policies and technologies on the continent.

5. Clean Energy Council

- Website: www.cleanenergycouncil.org.au

- Description: Leading clean energy organization in Australia, supporting the development of renewable technologies and the transition to a clean energy economy.

www.ingramcontent.com/pod-product-compliance
Lightning Source LLC
Chambersburg PA
CBHW071829210526
45479CB00001B/48